Lecture Notes in Mathematics

2051

For further volumes:
http://www.springer.com/series/304

FONDAZIONE
CIME
ROBERTO CONTI
CENTRO INTERNAZIONALE MATEMATICO ESTIVO
INTERNATIONAL MATHEMATICAL SUMMER CENTER

Fondazione C.I.M.E., Firenze

C.I.M.E. stands for *Centro Internazionale Matematico Estivo*, that is, International Mathematical Summer Centre. Conceived in the early fifties, it was born in 1954 in Florence, Italy, and welcomed by the world mathematical community: it continues successfully, year for year, to this day.

Many mathematicians from all over the world have been involved in a way or another in C.I.M.E.'s activities over the years. The main purpose and mode of functioning of the Centre may be summarised as follows: every year, during the summer, sessions on different themes from pure and applied mathematics are offered by application to mathematicians from all countries. A Session is generally based on three or four main courses given by specialists of international renown, plus a certain number of seminars, and is held in an attractive rural location in Italy.

The aim of a C.I.M.E. session is to bring to the attention of younger researchers the origins, development, and perspectives of some very active branch of mathematical research. The topics of the courses are generally of international resonance. The full immersion atmosphere of the courses and the daily exchange among participants are thus an initiation to international collaboration in mathematical research.

C.I.M.E. Director
Pietro ZECCA
Dipartimento di Energetica "S. Stecco"
Università di Firenze
Via S. Marta, 3
50139 Florence
Italy
e-mail: zecca@unifi.it

C.I.M.E. Secretary
Elvira MASCOLO
Dipartimento di Matematica "U. Dini"
Università di Firenze
viale G.B. Morgagni 67/A
50134 Florence
Italy
e-mail: mascolo@math.unifi.it

For more information see CIME's homepage: http://www.cime.unifi.it

CIME activity is carried out with the collaboration and financial support of:

- INdAM (Istituto Nazionale di Alta Matematica)

- MIUR (Ministero dell'Universitá e della Ricerca)

This course is partially supported by GDR-GDRE on CONTROLE DES EQUATIONS AUX DERIVEES PARTIELLES (CONEDP).

Vincent Rivasseau • Robert Seiringer
Jan Philip Solovej • Thomas Spencer

Quantum Many Body Systems

Cetraro, Italy 2010

Editors:
Alessandro Giuliani
Vieri Mastropietro
Jakob Yngvason

Springer

FONDAZIONE
CIME
ROBERTO CONTI

Vincent Rivasseau
Université Paris-Sud
Laboratoire de Physique Théorique
Orsay, France

Jan Philip Solovej
Department of Mathematics
University of Copenhagen
Copenhagen, Denmark

Robert Seiringer
McGill University
Department of Mathematics and Statistics
Montreal, QC, Canada

Thomas Spencer
Institute for Advanced Study
School of Mathematics
Princeton, NJ, USA

ISBN 978-3-642-29510-2 ISBN 978-3-642-29511-9 (eBook)
DOI 10.1007/978-3-642-29511-9
Springer Heidelberg New York Dordrecht London

Lecture Notes in Mathematics ISSN print edition: 0075-8434
ISSN electronic edition: 1617-9692

Library of Congress Control Number: 2012941621

Mathematics Subject Classification (2010): 82B10, 81V70, 82B28, 82B44

Printed on acid-free paper

Springer is part of Springer Science+Business Media (www.springer.com)

Preface

The idea that matter is composed by a huge number of particles (atoms) obeying simple laws of motion and that the macroscopic properties of bodies emerge as collective phenomena starting from this simple description dates back to Democritus (460–370 BC). Since then, physicists of all times have struggled with this fascinating idea in order to really understand the variety of our world starting from a simple microscopic description. Quantum many-body theory is nowadays the area of physics that provides the framework for understanding the collective behavior of vast assemblies of interacting quantum particles. Indeed, we know that matter is composed of atoms and molecules, interacting in accord with the laws of quantum mechanics. The equation describing the evolution and behavior of such assemblies, the Schrödinger equation with Coulomb forces, is well known: it can be written down and studied, and the properties of the system can in principle be understood at a qualitative and quantitative level, starting from this fundamental law. However, the number of variables required for describing the behavior of the microscopic constituents of any macroscopic body is enormous (it is of the order of the number of particles, which is $\sim 10^{23}$ for a mole of substance). Therefore, except for a few special but very important cases (e.g., ideal gases), the deduction of the macroscopic properties from the fundamental microscopic equation is a formidable mathematical challenge. This is, after all, not surprising: we know from experiments that complex and unexpected phenomena like superconductivity, superfluidity, and Bose–Einstein condensation, together with more familiar phenomena like magnetism or metallic behavior, emerge from the collective behavior of a huge number of quantum particles. These remarkable phenomena are a macroscopic manifestation of the law of quantum mechanics and cannot be understood by using Newtonian physics.

From the point of view of mathematical physicists, many-body theory provides an almost ideal field; the basic equation is known and one "just" needs to solve it in order to deduce and understand a number of interesting phenomena. Indeed, if the solutions of the Schrödinger equation with a large number of variables were known this would clearly be of extraordinary importance for physical applications as well as for technological developments. However, the mathematical difficulty of finding such solutions is enormous, but in past years a number of increasingly powerful and sophisticated techniques have been developed to extract relevant information without having to solve the equation itself. These methods include multiscale analysis, functional inequalities, localization estimates, cluster expansions, supersymmetry and stochastic and conformal geometry. The beautiful mathematics related to these developments has also proved to be useful as a bridge between different disciplines, ranging from algebraic geometry to the study of the shape of bird flocks. The interaction of mathematical and physical complexity has proved to be very fruitful for both fields.

The aim of the *CIME school* on *Quantum Many-Body Systems*, which took place in Cetraro (Italy) from August 30 to September 4, 2010, was to provide an introduction to the beautiful and powerful mathematical techniques developed in this field. The school was attended by 30 participants from several different countries, including Austria, Denmark, France, Germany, Italy, the UK, Ukraine and the USA. Although the school was primarily intended for graduate students, the interesting topics and the high reputation of the lecturers also attracted several more senior researchers.

The school consisted of four series of lectures, presented by V. Rivasseau, R. Seiringer, J.P. Solovej, and T. Spencer, and each series was organized into four lessons of 2 h each (two in the morning and one in the afternoon). In addition, one afternoon was devoted to short presentations of the research activity of the younger participants and one evening to a short description of the activities of some of the senior participants.

The lectures of Prof. Rivasseau gave an introduction to some results in solid-state physics, obtained via constructive Renormalization Group methods. While the focus was on the proof of Fermi liquid behavior for a system of nonrelativistic two-dimensional fermions above the BCS transition temperature one lecture was devoted to the exciting perspectives opened by the use of the same methods to quantum gravity.

The lectures of Prof. Seiringer were devoted to the mathematical physics of Bose gases and Bose–Einstein condensation and included a rigorous proof of the latter for dilute, interacting gases in the so-called Gross–Pitaevski limit. Starting from the basic notions he also discussed advanced topics like the analysis of rotating traps and the emergence of lattices of quantized vortices.

Prof. Solovej provided a comprehensive introduction to quantum Coulomb systems. He gave a self-contained presentation of the functional analytic methods used to prove thermodynamic stability of coulombic matter, following a recently developed approach that allows to treat, on the same footing, translation and non-translation invariant systems of charged fermions and bosons.

Finally, Prof. Spencer described the rigorous and powerful methods of supersymmetry and their application to the problem of the localization–delocalization transition in the Anderson model and in random matrices. Moreover, he gave a tutorial review of some classical results and techniques, such as the use of Ward Identities in the XY model.

The atmosphere at the school was very lively, many questions and comments arose during and after each lecture, and scientific discussions took place; the students profited very much from the possibility of close interactions with the lecturers.

As Editors of these Lectures Notes we would like to thank the people and institutions who contributed to the success of the course. In particular, it is our pleasure to thank the Scientific Committee of CIME for supporting our project; the Director, Prof. Pietro Zecca, and the Secretary, Prof. Elvira Mascolo, for their support during the organization. Special thanks go to the lecturers, who offered a unique occasion to the participants to enter this beautiful field.

Roma, Italy — Alessandro Giuliani
Roma, Italy — Vieri Mastropietro
Vienna, Austria — Jakob Yngvason

Acknowledgements

We gratefully acknowledge financial support from the ERC Starting Grant CoMBoS-239694, from the CIME foundation and from the International Association in Mathematical Physics.

Acknowledgments

Contents

Chapter 1
Introduction to the Renormalization Group with Applications to Non-relativistic Quantum Electron Gases

Vincent Rivasseau

1.1 Introduction to QFT and Renormalization

In these lectures we review the rigorous work on many Fermions models which led to the first constructions of interacting Fermi liquids in two dimensions, and to the proof that they obey different scaling regimes depending on the shape of the Fermi surface. We also review progress on the three dimensional case.

We start with a pedagogical introduction on quantum field theory and perturbative renormalization. Emphasis is then put on using renormalization around the Fermi surface in a constructive way, in which all orders of perturbation theory are summed rigorously.

1.1.1 Gaussian Measures

A finite dimensional centered normalized Gaussian measure $d\mu_C$ is defined through its covariance. Consider a finite dimensional space $\mathbb{R}^N$ and a symmetric positive definite N by N matrix A. The inverse of the matrix A is also a definite positive symmetric N by N matrix $C = A^{-1}$ called the covariance associated to A. The corresponding centered normalized Gaussian measure is

$$d\mu_C = (2\pi)^{-N/2}\sqrt{\det A}\ \ e^{-\frac{1}{2}{}^t XAX} d^N X, \tag{1.1}$$

so that $\int d\mu_C = 1$.

V. Rivasseau (✉)
Laboratoire de physique théorique, Université Paris-Sud, 91405 Orsay, France
e-mail: vincent.rivasseau@gmail.com

V. Rivasseau et al., *Quantum Many Body Systems*, Lecture Notes in Mathematics 2051,
DOI 10.1007/978-3-642-29511-9_1,

To understand $d\mu_C$ it is better to know C than A since the moments or correlation functions of a Gaussian measure can be expressed simply as sums of monomials in C. In fact formula (1.1) perfectly makes sense if C is non invertible, and even for $C = 0$; but the corresponding measure has no density with respect to the Lebesgue measure in this case (for $C = 0$ $d\mu_C$ is just Dirac's δ function at the origin). The reader familiar with eg ordinary linear PDE's knows that the essential point is to invert the matrix or the operator, hence to know the "Green's function". But quadratic forms have linear equations as their variational solutions, so both problems are linked.

Sine any function can be approximated by polynomials, probability measures are characterized by their moments, that is by the integrals they return for each polynomial of the integration variables.

The theorem which computes the moments of a Gaussian measure in terms of the covariance is fundamental in QFT and known there under the name of Wick's theorem. It expresses the result as the sum over all possible pairings or the variables of a product of covariances between the paired variables:

$$\int X_{i_1} \dots X_{i_n} d\mu_C = \sum_G \prod_{\ell \in G} C_{i_{b(\ell)}, i_{e(\ell)}}, \tag{1.2}$$

where G runs over all Wick contractions or pairings of the labels $1, \dots, n$. Each pair ℓ is pictured as a line a pair of labels $b(\ell)$ and $e(\ell)$.[1]

The theory of Gaussian measures and Wick's theorem extends to infinite dimensional spaces, in which case the covariance C usually becomes a positive kernel $C(x, y)$ in a distribution space. Such a kernel can be considered also as an operator acting on functions through $C.f = \int C(x, y) f(y)\, dy$. The Dirac kernel $C(x, y) = \delta(x - y)$ corresponds to the identity operator: $Cf = f$. When C is positive definite and obeys some regularity properties it is the covariance of a truly well defined Gaussian measure in some infinite dimensional space of distributions, through an extension of Bochner's theorem known as Minlos Theorem; see [1] for details.

1.1.2 Functional Integrals

In QFT, like in grand-canonical statistical mechanics, particle number is not conserved. Cross sections in scattering experiments contain the physical information

[1]If we order the pair $(b(\ell), e(\ell))$, b can be considered the "beginning" and e the "end" of the line, which is equivalent to orient the line. However the ordinary real-variable Wick's theorem does not require such an orientation, which becomes necessary only for complex or Grassmann Gaussian measures.

of the theory.[2] They are the matrix elements of the diffusion matrix $\mathscr{S}$. Under suitable conditions they are expressed in terms of the Green functions G_N of the theory through so-called "reduction formulae".

Green functions are time ordered vacuum expectation values of the field ϕ, which is operator valued and acts on the Fock space:

$$G_N(z_1, \ldots, z_N) = <\psi_0, T[\phi(z_1)\ldots\phi(z_N)]\psi_0> . \tag{1.3}$$

Here ψ_0 is the vacuum state and the T-product orders $\phi(z_1)\ldots\phi(z_N)$ according to increasing times.

Consider a Lagrangian field theory, and split the total Lagrangian as the sum of a free plus an interacting piece, $\mathscr{L} = \mathscr{L}_0 + \mathscr{L}_{int}$. The Gell-Mann–Low formula expresses the Green functions as vacuum expectation values of a similar product of free fields with an $e^{i\int \mathscr{L}_{int}}$ insertion:

$$G_N(z_1, \ldots, z_N) = \frac{<\psi_0, T\left[\phi(z_1)\ldots\phi(z_N)e^{i\int dx \mathscr{L}_{int}(\phi(x))}\right]\psi_0>}{<\psi_0, T(e^{i\int dx \mathscr{L}_{int}(\phi(x))})\psi_0>}. \tag{1.4}$$

In the functional integral formalism proposed by Feynman [2], the Gell-Mann–Low formula is replaced by a functional integral in terms of an (ill-defined) "integral over histories" which is formally the product of Lebesgue measures over all space time. The corresponding formula is the Feynman–Kac formula:

$$G_N(z_1, \ldots, z_N) = \frac{\int \prod_j \phi(z_j) e^{i\int \mathscr{L}(\phi(x))dx} D\phi}{\int e^{i\int \mathscr{L}(\phi(x))dx} D\phi}. \tag{1.5}$$

The integrand in (1.5) contains now the full Lagrangian $\mathscr{L} = \mathscr{L}_0 + \mathscr{L}_{int}$ instead of the interacting one. This is interesting to expose symmetries of the theory which may not be separate symmetries of the free and interacting Lagrangians, for instance gauge symmetries. Perturbation theory and the Feynman rules can still be derived as explained in the next subsection. But (1.5) is also well adapted to constrained quantization and to the study of non-perturbative effects.

For general references on QFT, see [3–5].

1.1.3 Statistical Mechanics and Thermodynamic Quantities

There is a deep analogy between the Feynman–Kac formula and the formula which expresses correlation functions in classical statistical mechanics.

[2]Correlation functions play this fundamental role in statistical mechanics.

The partition function of a statistical mechanics grand canonical ensemble described by a Hamiltonian H at temperature T and chemical potential μ in volume Λ is

$$Z_\Lambda = Tr\, e^{-\beta(H-\mu N)}, \tag{1.6}$$

where $\beta = 1/kT$, and the trace may be either a classical integration in the phase-space corresponding to volume Λ or in the quantum case a trace on the relevant Hilbert space (Fock space).

The main problem in *equilibrium statistical mechanics* is to compute the Gibbs states or thermodynamic limits of a finite size statistical system in the limit of its size becoming arbitrarily large. More precisely the main issue is to compute the limit as the volume Λ of the system tends to infinity of the *logarithm* of the partition function per unit volume. Indeed intensive thermodynamic quantities such as the pressure, free energy, entropy or heat capacity can be derived from it. In the simplest case where there is a unique Gibbs state or phase for the system (this is usually true at sufficiently high temperature) the thermodynamic limit does not depend on the way the volume is sent to infinity, provided it is done in a sufficiently regular way, nor of the boundary conditions.

Hence the main mathematical issue is to derive simple tools to compute, for a d dimensional system

$$p = \lim_{\Lambda \to \mathbb{R}^d} \frac{1}{|\Lambda|} \log Z_\Lambda. \tag{1.7}$$

More precisely all the detailed information on the Gibbs states is encoded in the list of their *correlation functions*, which are derivatives of the logarithm of the partition function with respect to appropriate sources. For instance for a lattice Ising model the partition function is

$$Z_\Lambda = \sum_{\{\sigma_x = \pm 1\}} e^{-L(\sigma)} \tag{1.8}$$

and the correlation functions are

$$\langle \prod_{i=1}^{n} \sigma_{x_i} \rangle = \frac{\sum_{\{\sigma_x = \pm 1\}} e^{-L(\sigma)} \prod_i \sigma_{x_i}}{\sum_{\{\sigma_x = \pm 1\}} e^{-L(\sigma)}}, \tag{1.9}$$

where x labels the discrete sites of the lattice. The sum is over configurations $\{\sigma_x = \pm 1\}$ which associate a "spin" with value $+1$ or -1 to each such site and $L(\sigma)$ contains usually nearest neighbor interactions and possibly a magnetic field h:

$$L(\sigma) = \sum_{x,y \text{ nearest neighbors}} J \sigma_x \sigma_y + \sum_x h \sigma_x. \tag{1.10}$$

By analytically continuing (1.5) to imaginary time, or Euclidean space, it is possible to complete the analogy with (1.9), hence to establish a firm contact between Euclidean QFT and statistical mechanics [6–8].

1.1.4 Schwinger Functions

This idea also allows to give much better meaning to the path integral, at least for a free Bosonic field. Indeed the free Euclidean measure can be defined easily as a Gaussian measure, because in Euclidean space L_0 is a quadratic form of positive type.[3]

The Green functions continued to Euclidean points are called the Schwinger functions of the model, and are given by the Euclidean Feynman–Kac formula:

$$S_N(z_1,\ldots,z_N) = Z^{-1}\int \prod_{j=1}^{N} \phi(z_j) e^{-\int \mathcal{L}(\phi(x))dx} D\phi, \tag{1.11}$$

$$Z = \int e^{-\int \mathcal{L}(\phi(x))dx} D\phi. \tag{1.12}$$

The simplest interacting field theory is the theory of a one component scalar Bosonic field ϕ with quartic interaction $\lambda\phi^4$ (ϕ^3, which is simpler, is unstable). In $\mathbb{R}^d$ it is called the ϕ^4_d model. For $d = 2, 3$ this model is super-renormalizable and has been built non-perturbatively by constructive field theory (see [1, 9]). In these dimensions the model is unambiguously related to its perturbation series [10, 11] through Borel summability [12]. For $d = 4$ the model is just renormalizable, and provides the simplest pedagogical introduction to perturbative renormalization theory. But because of the Landau ghost or triviality problem explained in Sect. 1.1.11, the model presumably does not exist as a true interacting theory at the non perturbative level (see [9] for a discussion of this subtle issue).

Formally the Schwinger functions of ϕ^4_d are the moments of the measure:

$$d\nu = \frac{1}{Z} e^{-\frac{\lambda}{4!}\int\phi^4-(m^2/2)\int\phi^2-(a/2)\int(\partial_\mu\phi\partial^\mu\phi)} D\phi, \tag{1.13}$$

where

[3] However the functional space that supports this measure is not in general a space of smooth functions, but rather of distributions. This was already true for functional integrals such as those of Brownian motion, which are supported by continuous but not differentiable paths. Therefore "functional integrals" in quantum field theory should more appropriately be called "distributional integrals".

- λ is the coupling constant, usually assumed positive or complex with positive real part; remark the arbitrary but convenient 1/4! factor to take into account the symmetry of permutation of all fields at a local vertex.
- m is the mass, which fixes an energy scale for the theory.
- a is the wave function constant. It can be set to 1 by a rescaling of the field.
- Z is a normalization factor which makes (1.13) a probability measure.
- $D\phi$ is a formal (mathematically ill-defined) product $\prod_{x \in \mathbb{R}^d} d\phi(x)$ of Lebesgue measures at every point of $\mathbb{R}^d$.

The Gaussian part of the measure is

$$d\mu(\phi) = \frac{1}{Z_0} e^{-(m^2/2)\int \phi^2 - (a/2)\int (\partial_\mu \phi \partial^\mu \phi)} D\phi. \tag{1.14}$$

where Z_0 is again the normalization factor which makes (1.14) a probability measure.

More precisely if we consider the translation invariant propagator $C(x, y) \equiv C(x - y)$ (with slight abuse of notation), whose Fourier transform is

$$C(p) = \frac{1}{(2\pi)^d} \frac{1}{p^2 + m^2}, \tag{1.15}$$

we can use Minlos theorem and the general theory of Gaussian processes to define $d\mu(\phi)$ as the centered Gaussian measure on the Schwartz space of tempered distributions $S'(\mathbb{R}^d)$ whose covariance is C. A Gaussian measure is uniquely defined by its moments, or the integral of polynomials of fields. Explicitly this integral is zero for a monomial of odd degree, and for even $n = 2p$ it is equal to

$$\int \phi(x_1) \ldots \phi(x_n) d\mu(\phi) = \sum_{\mathcal{W}} \prod_{\ell \in \mathcal{W}} C(x_{b(\ell)}, x_{e(\ell)}), \tag{1.16}$$

where the sum runs over all the $2p!! = (2p-1)(2p-3)\ldots 5.3.1$ Wick pairings $\mathcal{W}$ of the $2p$ arguments into the p disjoint pairs $\ell = (b(\ell), e(\ell))$.

Note that since for $d \geq 2$, $C(p)$ is not integrable, $C(x, y)$ must be understood as a distribution. It is therefore convenient to also use regularized kernels, for instance

$$C_\kappa(p) = \frac{1}{(2\pi)^d} \frac{e^{-\kappa(p^2+m^2)}}{p^2 + m^2} = \int_\kappa^\infty e^{-\alpha(p^2+m^2)} d\alpha \tag{1.17}$$

whose Fourier transform $C_\kappa(x, y)$ is a smooth function and not a distribution:

$$C_\kappa(x, y) = \int_\kappa^\infty e^{-\alpha m^2 - (x-y)^2/4\alpha} \frac{d\alpha}{\alpha^{D/2}}. \tag{1.18}$$

$\alpha^{-D/2} e^{-(x-y)^2/4\alpha}$ is the *heat kernel*. Therefore this α-representation has also an interpretation in terms of Brownian motion:

$$C_\kappa(x,y) = \int_\kappa^\infty d\alpha \exp(-m^2\alpha)\, P(x,y;\alpha) \tag{1.19}$$

where $P(x,y;\alpha) = (4\pi\alpha)^{-d/2}\exp(-|x-y|^2/4\alpha)$ is the Gaussian probability distribution of a Brownian path going from x to y in time α.

Such a regulator κ is called an ultraviolet cutoff, and we have (in the distribution sense) $\lim_{\kappa\to 0} C_\kappa(x,y) = C(x,y)$. Remark that due to the non zero m^2 mass term, the kernel $C_\kappa(x,y)$ decays exponentially at large $|x-y|$ with rate m. For some constant K and $d > 2$ we have:

$$|C_\kappa(x,y)| \le K\kappa^{1-d/2} e^{-m|x-y|}. \tag{1.20}$$

It is a standard useful construction to build from the Schwinger functions the connected Schwinger functions, given by:

$$C_N(z_1,\dots,z_N) = \sum_{P_1\cup\ldots\cup P_k=\{1,\ldots,N\};\; P_i\cap P_j=0} (-1)^{k+1}\prod_{i=1}^{k} S_{p_i}(z_{j_1},\dots,z_{j_{p_i}}), \tag{1.21}$$

where the sum is performed over all distinct partitions of $\{1,\dots,N\}$ into k subsets $P_1,\dots,P_k$, P_i being made of p_i elements called $j_1,\dots,j_{p_i}$. For instance in the ϕ^4 theory, where all odd Schwinger functions vanish due to the unbroken $\phi \to -\phi$ symmetry, the connected 4-point function is simply:

$$\begin{aligned} C_4(z_1,\dots,z_4) = {} & S_4(z_1,\dots,z_4) - S_2(z_1,z_2)S_2(z_3,z_4) \\ & -S_2(z_1,z_3)S_2(z_2,z_4) - S_2(z_1,z_4)S_2(z_2,z_3). \end{aligned} \tag{1.22}$$

1.1.5 Feynman Graphs

The full interacting measure may now be defined as the multiplication of the Gaussian measure $d\mu(\phi)$ by the interaction factor:

$$d\nu = \frac{1}{Z} e^{-\frac{\lambda}{4!}\int \phi^4(x)dx} d\mu(\phi) \tag{1.23}$$

and the Schwinger functions are the normalized moments of this measure:

$$S_N(z_1,\dots,z_N) = \int \phi(z_1)\dots\phi(z_N)d\nu(\phi). \tag{1.24}$$

Expanding the exponential as a power series in the coupling constant λ, one obtains a formal expansion for the Schwinger functions:

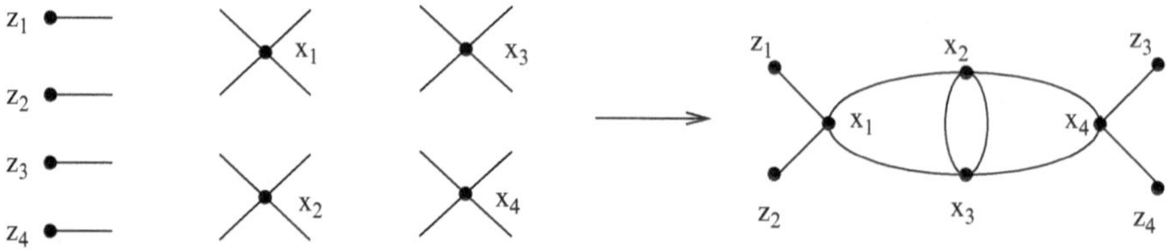

Fig. 1.1 A possible contraction scheme with $n = N = 4$

$$S_N(z_1,\ldots,z_N) = \frac{1}{Z}\sum_{n=0}^{\infty}\frac{(-\lambda)^n}{n!}\int [\int \frac{\phi^4(x)dx}{4!}]^n \phi(z_1)\ldots\phi(z_N)d\mu(\phi). \quad (1.25)$$

It is now possible to perform explicitly the functional integral of the corresponding polynomial. The result is at any order n a sum over $(4n+N-1)!!$ Wick contractions schemes $\mathcal{W}$, i.e. over all the ways of pairing together $4n+N$ fields into $2n+N/2$ pairs. The weight or *amplitude* of such a scheme $\mathcal{W}$ is the spatial integral over $x_1,\ldots,x_n$ of the integrand $\prod_{\ell\in\mathcal{W}} C(x_{i_{b(\ell)}}, x_{i\,e(\ell)})$ times the factor $\frac{1}{n!}(\frac{-\lambda}{4!})^n$. Such amplitudes are functions (in fact distributions) of the external positions $z_1,\ldots,z_N$. They may diverge either because they are integrals over all of $\mathbb{R}^4$ (no volume cutoff) or because the integrand is typically unbounded due to the singularities in the propagator C at coinciding points.

Labeling the n dummy integration variables in (1.25) as $x_1,\ldots,x_n$, we draw a line ℓ for each contraction of two fields. Each position $x_1,\ldots,x_n$ is then associated to a four-legged vertex and each external source z_i to a one-legged vertex, as shown in Fig. 1.1.

It is convenient to draw these Wick contractions and to regroup all contractions which give rise to the same drawing or graph. There are some subtleties about labels.

Example 1.1. For the normalization at order 1 we have 4 fields, hence 3 Wick contractions, which all give the same graph. For the 2 point function at order 1 we have 6 fields, and 15 Wick contractions which fall into 2 categories with weight 3 and 12.

Here are some additional observations.

- The great advantage of Feynman graphs is that they form a combinatoric species in the sense of Joyal [13] whose logarithm can be computed as the species of connected graphs. As we already remarked, the computation of this logarithm is a key physical problem.
- However Feynman graphs *proliferate*, i.e. their generating functional $\sum_n \frac{a_n}{n!}\lambda^n$ has zero radius of convergence in λ. At the heart of any constructive strategy [1, 9, 14–18], lies the replacement of the proliferating species of Feynman graphs by a better one [19], typically the species of forests. The corresponding connected species is the species of trees, which does not proliferate. Indeed by Cayley's theorem there are only n^{n-2} labeled trees on n vertices. This is why

constructive expansions converge while ordinary perturbative expansions dont. Constructive theory ultimately may be considered just as repacking Feynman graphs in some clever way according to underlying forests [20]. See also the discussion in Sect. 1.1.16 and below.

- The computation factorizes nicely into the connected components of the graphs. These components may or may not have external arguments. In the expansion for the normalized functions the *vacuum* components (i.e. those without external arguments) factor out and disappear. Only graphs whose connected components all contain external arguments remain.
- If we further search for elementary bricks of the expansion, we can consider the *connected* Schwinger functions like (1.22). In the expansion of these functions only the graphs with a single connected component containing all external arguments survive.

1.1.6 Feynman Rules

The "Feynman rules" summarize how to compute the amplitude associated to a Feynman graph with its correct combinatoric factor.

We always use the following notations for a graph G:

- $n(G)$ or simply n is the number of internal vertices of G, or the order of the graph.
- $l(G)$ or l is the number of internal lines of G, i.e. lines hooked at both ends to an internal vertex of G.
- $N(G)$ or N is the number of external vertices of G; it corresponds to the order of the Schwinger function one is looking at. When $N = 0$ the graph is a vacuum graph, otherwise it is called an N-point graph.
- $c(G)$ or c is the number of connected components of G.
- $L(G)$ or L is the number of independent loops of G.

For a *regular* ϕ^4 graph, i.e. a graph which has no line hooked at both ends to external vertices, we have the relations:

$$l(G) = 2n(G) - N(G)/2, \tag{1.26}$$

$$L(G) = l(G) - n(G) + c(G) = n(G) + 1 - N(G)/2. \tag{1.27}$$

where in the last equality we assume connectedness of G, hence $c(G) = 1$.

A *subgraph* F of a graph G is a subset of internal lines of G, together with the corresponding attached vertices. Lines in the subset defining F are the internal lines of F, and their number is simply $l(F)$, as before. Similarly all the vertices of G hooked to at least one of these internal lines of F are called the internal vertices of F and considered to be in F; their number by definition is $n(F)$. Finally a good convention is to call external half-line of F every half-line of G which is not in F

but which is hooked to a vertex of F; it is then the number of such external half-lines which we call $N(F)$. With these conventions one has for ϕ^4 subgraphs the same relation (1.26) as for regular ϕ^4 graphs.

To compute the amplitude associated to a ϕ^4 graph, we have to add the contributions of the corresponding contraction schemes. This is summarized by the "Feynman rules":

- To each line ℓ with end vertices at positions x_ℓ and y_ℓ, associate a propagator $C(x_\ell, y_\ell)$.
- To each internal vertex, associate $(-\lambda)/4!$.
- Count all the contraction schemes giving this diagram. The number should be of the form $(4!)^n n!/S(G)$ where $S(G)$ is an integer called the symmetry factor of the diagram. The 4! represents the permutation of the fields hooked to an internal vertex.
- Multiply all these factors, divide by $n!$ and sum over the position of all internal vertices.

The formula for the bare amplitude of a graph is therefore, as a distribution in $z_1, \ldots .z_N$:

$$A_G(z_1, \ldots, z_N) \equiv \int \prod_{i=1}^{n} dx_i \prod_{\ell \in G} C(x_{i_{b(\ell)}}, x_{i_{e(\ell)}}). \tag{1.28}$$

This is the "direct" or "x-space" representation of a Feynman integral. As stated above, this integral suffers of possible divergences. But the corresponding quantity with both volume cutoff Λ and ultraviolet cutoff κ, namely

$$A^{\kappa}_{G,\Lambda}(z_1, \ldots, z_N) \equiv \int_{\Lambda^n} \prod_{i=1}^{n} dx_i \prod_{\ell \in G} C_\kappa(x_{i_{b(\ell)}}, x_{i_{e(\ell)}}), \tag{1.29}$$

is well defined. The integrand is indeed bounded and the integration domain Λ is assumed compact.

The *unnormalized* Schwinger functions are therefore formally given by the sum over all graphs with the right number of external lines of the corresponding Feynman amplitudes:

$$ZS_N = \sum_{\phi^4 \text{ graphs } G \text{ with } N(G)=N} \frac{(-\lambda)^{n(G)}}{S(G)} A_G. \tag{1.30}$$

Z itself, the normalization, is given by the sum of all vacuum amplitudes:

$$Z = \sum_{\phi^4 \text{ graphs } G \text{ with } N(G)=0} \frac{(-\lambda)^{n(G)}}{S(G)} A_G. \tag{1.31}$$

We already remarked that the species of Feynman graphs *proliferate* at large orders. More precisely the total number of ϕ^4 Feynman graphs at order n with N

external arguments is $(4n + N)!!$. Taking into account Stirling's formula and the symmetry factor $1/n!$ from the exponential we expect perturbation theory at large order to behave as $K^n n!$ for some constant K. Indeed at order n the amplitude of a Feynman graph is a 4n-dimensional integral. It is reasonable to expect that in average it should behave as c^n for some constant c. But this means that one should expect zero radius of convergence for the series (1.30). This is not too surprising. Even the one-dimensional integral

$$F(g) = \int_{-\infty}^{+\infty} e^{-x^2/2 - \lambda x^4/4!} dx \tag{1.32}$$

is well-defined only for $\lambda \geq 0$. We cannot hope infinite dimensional functional integrals of the same kind to behave better than this one dimensional integral. In mathematically precise terms, F is not analytic near $\lambda = 0$, but only Borel summable. Borel summability [12] is therefore the best we can hope for the ϕ^4 theory, and we mentioned that it has indeed been established for the ϕ^4 theory in dimensions 2 and 3 [10, 11].

From translation invariance, we do not expect $A^{\kappa}_{G,\Lambda}$ to have a limit as $\Lambda \to \infty$ if there are vacuum subgraphs in G. But obviously an amplitude factorizes as the product of the amplitudes of its connected components.

With simple combinatoric verification at the level of contraction schemes we can factorize the sum over all vacuum graphs in the expansion of unnormalized Schwinger functions, hence get for the normalized functions a formula analog to (1.30):

$$S_N = \sum_{\substack{\phi^4 \text{ graphs } G \text{ with } N(G)=N \\ G \text{ without any vacuum subgraph}}} \frac{(-\lambda)^{n(G)}}{S(G)} A_G. \tag{1.33}$$

Now in (1.33) it is possible to pass to the thermodynamic limit (in the sense of formal power series) because using the exponential decrease of the propagator, each individual graph has a limit at fixed external arguments. There is of course no need to divide by the volume for that because each connected component in (1.33) is tied to at least one external source, and they provide the necessary breaking of translation invariance.

Finally one can find the perturbative expansions for the connected Schwinger functions and the vertex functions. As expected, the connected Schwinger functions are given by sums over connected amplitudes:

$$C_N = \sum_{\phi^4 \text{ connected graphs } G \text{ with } N(G)=N} \frac{(-\lambda)^{n(G)}}{S(G)} A_G \tag{1.34}$$

and the vertex functions are the sums of the *amputated* amplitudes for proper graphs, also called one-particle-irreducible. They are the graphs which remain connected even after removal of any given internal line. The amputated amplitudes are defined in momentum space by omitting the Fourier transform of the propagators of the external lines. It is therefore convenient to write these amplitudes in the so-called momentum representation:

$$\Gamma_N(z_1,\dots,z_N) = \sum_{\phi^4 \text{ proper graphs } G \text{ with } N(G)=N} \frac{(-\lambda)^{n(G)}}{S(G)} A_G^T(z_1,\dots,z_N), \quad (1.35)$$

$$A_G^T(z_1,\dots,z_N) \equiv \frac{1}{(2\pi)^{dN/2}} \int dp_1 \dots dp_N e^{i\sum p_i z_i} A_G(p_1,\dots,p_N), \quad (1.36)$$

$$A_G(p_1,\dots,p_N) = \int \prod_{\ell \text{ internal line of } G} \frac{d^d p_\ell}{p_\ell^2+m^2} \prod_{v\in G} \delta(\sum_\ell \epsilon(v,\ell)\, p_\ell). \quad (1.37)$$

Remark in (1.37) the δ functions which ensure momentum conservation at each internal vertex v; the sum inside is over both internal and external momenta; each internal line is oriented in an arbitrary way (from $b(\ell)$ to $e(\ell)$) and each external line is oriented towards the inside of the graph. The incidence matrix $\epsilon(v,\ell)$ captures in a nice way the information on the internal lines.[4] It is 1 if the line ℓ arrives at v, -1 if it starts from v and 0 otherwise. Remark also that there is an overall momentum conservation rule $\delta(p_1+\ldots+p_N)$ hidden in (1.37). The drawback of the momentum representation lies in the necessity for practical computations to eliminate the δ functions by a "momentum routing" prescription, and there is no canonical choice for that. Although this is rarely explicitly explained in the quantum field theory literature, such a choice of a momentum routing is equivalent to the choice of a particular spanning tree of the graph.

1.1.7 Scale Analysis and Renormalization

In order to analyze the ultraviolet or short distance limit according to the renormalization group method [22], we can cut the propagator C into slices C_i so that $C = \sum_{i=0}^{\infty} C_i$. This can be done conveniently within the parametric representation, since α in this representation roughly corresponds to $1/p^2$. So we can define the propagator within a slice as

$$C_0 = \int_1^\infty e^{-m^2\alpha - \frac{|x-y|^2}{4\alpha}} \frac{d\alpha}{\alpha^{d/2}}, \quad C_i = \int_{M^{-2i}}^{M^{-2(i-1)}} e^{-m^2\alpha - \frac{|x-y|^2}{4\alpha}} \frac{d\alpha}{\alpha^{d/2}} \quad \text{for } i \geq 1. \quad (1.38)$$

where M is a fixed number, for instance 10, or 2, or e. We can intuitively imagine C_i as the piece of the field oscillating with Fourier momenta essentially of size M^i.

[4] Strictly speaking this is true only for semi-regular graphs, i.e. graphs without tadpoles, i.e. without lines which start and end at the same vertex, see [21].

In fact it is easy to prove the bound (for $d > 2$)

$$|C_i(x,y)| \leq K.M^{(d-2)i} e^{-M^i |x-y|} \tag{1.39}$$

where K is some constant.

Now the full propagator with ultraviolet cutoff M^ρ, ρ being a large integer, may be viewed as a sum of slices:

$$C_{\leq \rho} = \sum_{i=0}^{\rho} C_i. \tag{1.40}$$

Then the basic renormalization group step is made of two main operations:

- A functional integration
- The computation of a logarithm

Indeed decomposing a covariance in a Gaussian process corresponds to a decomposition of the field into independent Gaussian random variables ϕ^i, each distributed with a measure $d\mu_i$ of covariance C_i. Let us introduce

$$\Phi_i = \sum_{j=0}^{i} \phi_j. \tag{1.41}$$

This is the "low-momentum" field for all frequencies lower than i. The RG idea is that starting from scale ρ and performing $\rho - i$ steps, one arrives at an effective action for the remaining field Φ_i. Then, writing $\Phi_i = \phi_i + \Phi_{i-1}$, one splits the field into a "fluctuation" field ϕ_i and a "background" field Φ_{i-1}. The first step, functional integration, is performed solely on the fluctuation field, so it computes

$$Z_{i-1}(\Phi_{i-1}) = \int d\mu_i(\phi_i) e^{-S_i(\phi_i + \Phi_{i-1})}. \tag{1.42}$$

Then the second step rewrites this quantity as the exponential of an effective action, hence simply computes

$$S_{i-1}(\Phi_{i-1}) = -\log[Z_{i-1}(\Phi_{i-1})]. \tag{1.43}$$

Now $Z_{i-1} = e^{-S_{i-1}}$ and one can iterate! The flow from the initial bare action $S = S_\rho$ for the full field to an effective renormalized action S_0 for the last "slowly varying" component ϕ_0 of the field is similar to the flow of a dynamical system. Its evolution is decomposed into a sequence of discrete steps from S_i to S_{i-1}.

This renormalization group strategy can be best understood on the system of Feynman graphs which represent the perturbative expansion of the theory. The first step, functional integration over fluctuation fields, means that we have to consider subgraphs with all their internal lines in higher slices than any of their external lines. The second step, taking the logarithm, means that we have to consider

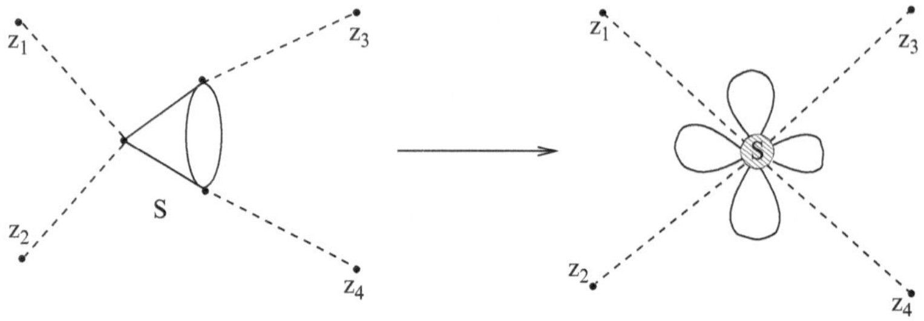

Fig. 1.2 A high energy subgraph **S** seen from lower energies looks quasi-local

only *connected* such subgraphs. We call such connected subgraphs *quasi-local.* Renormalizability is then a non trivial result that combines locality and power counting for these quasi-local subgraphs.

1.1.8 Locality, Power Counting

Locality simply means that *quasi-local* subgraphs S look *local* when seen through their external lines. Indeed since they are connected and since their internal lines have scale say $\geq i$, all the internal vertices are roughly at distance M^{-i}. But the external lines have scales $\leq i-1$, which only distinguish details larger than $M^{-(i-1)}$. Therefore they cannot distinguish the internal vertices of S one from the other. Hence quasi-local subgraphs look like "fat dots" when seen through their external lines, see Fig. 1.2. Obviously this locality principle is completely independent of dimension.

Power counting is a rough estimate which compares the size of a fat dot such as S in Fig. 1.2 with N external legs to the coupling constant that would be in front of an *exactly local* $\int \phi^N(x)dx$ interaction term if it were in the Lagrangian. To simplify we now assume that the internal scales are all equal to i, the external scales are $O(1)$, and we do not care about constants and so on, but only about the dependence in i as i gets large. We must first save one internal position such as the barycenter of the fat dot or the position of a particular internal vertex to represent the $\int dx$ integration in $\int \phi^N(x)dx$. Then we must integrate over the positions of all internal vertices of the subgraph *save that one*. This brings about a weight $M^{-di(n-1)}$, because since S is connected we can use the decay of the internal lines to evaluate these $n-1$ integrals. Finally we should not forget the prefactor $M^{(D-2)li}$ coming from (1.39), for the l internal lines. Multiplying these two factors and using relation (1.26)–(1.27) we obtain that the "coupling constant" or factor in front of the fat dot is of order $M^{-di(n-1)+2i(2n-N/2)} = M^{\omega(G)}$, if we define the superficial degree of divergence of a ϕ^4_d connected graph as:

$$\omega(G) = (d-4)n(G) + d - \frac{d-2}{2}N(G). \tag{1.44}$$

So power counting, in contrast with locality, depends on the space-time dimension.

Let us return to the concrete example of Fig. 1.2. A 4-point subgraph made of three vertices and four internal lines at a high slice i index. If we suppose the four external dashed lines have much lower index, say of order unity, the subgraph looks almost local, like a fat dot at this unit scale. We have to save one vertex integration for the position of the fat dot. Hence the coupling constant of this fat dot is made of two vertex integrations and the four weights of the internal lines (in order not to forget these internal line factors we kept internal lines apparent as four tadpoles attached to the fat dot in the right of Fig. 1.2). In dimension 4 this total weight turns out to be independent of the scale.

1.1.9 Renormalization, Effective Constants

At lower scales propagators can branch either through the initial bare coupling or through any such fat dot in all possible ways because of the combinatorial rules of functional integration. Hence they feel effectively a new coupling which is the sum of the bare coupling plus all the fat dot corrections coming from higher scales. To compute these new couplings only graphs with $\omega(G) \geq 0$, which are called primitively divergent, really matter because their weight does not decrease as the gap i increases.

- If $d = 2$, we find $\omega(G) = 2 - 2n$, so the only primitively divergent graphs have $n = 1$, and $N = 0$ or $N = 2$. The only divergence is due to the "tadpole" loop $\int \frac{d^2 p}{(p^2+m^2)}$ which is logarithmically divergent.
- If $d = 3$, we find $\omega(G) = 3 - n - N/2$, so the only primitively divergent graphs have $n \leq 3$, $N = 0$, or $n \leq 2$ and $N = 2$. Such a theory with only a finite number of "primitively divergent" subgraphs is called super-renormalizable.
- If $d = 4$, $\omega(G) = 4 - N$. Every two point graph is quadratically divergent and every four point graph is logarithmically divergent. This is in agreement with the superficial degree of these graphs being respectively 2 and 0. The couplings that do not decay with i all correspond to terms that were already present in the Lagrangian, namely $\int \phi^4$, $\int \phi^2$ and $\int (\nabla\phi).(\nabla\phi)$.[5] Hence the structure of the Lagrangian resists under change of scale, although the values of the coefficients can change. The theory is called just renormalizable.
- Finally for $d > 4$ we have infinitely many primitively divergent graphs with arbitrarily large number of external legs, and the theory is called non-renormalizable, because fat dots with N larger than 4 are important and they correspond to new

[5] Because the graphs with $N = 2$ are quadratically divergent we must Taylor expand the quasi local fat dots until we get convergent effects. Using parity and rotational symmetry, this generates only a logarithmically divergent $\int (\nabla\phi).(\nabla\phi)$ term beyond the quadratically divergent $\int \phi^2$. Furthermore this term starts only at $n = 2$ or two loops, because the first tadpole graph at $N = 2$, $n = 1$ is *exactly* local.

couplings generated by the renormalization group which are not present in the initial bare Lagrangian.

To summarize:

- Locality means that quasi-local subgraphs look local when seen through their external lines. It holds in any dimension.
- Power counting gives the rough size of the new couplings associated to these subgraphs as a function of their number N of external legs, of their order n and of the dimension of space time d.
- Renormalizability (in the ultraviolet regime) holds if the structure of the Lagrangian resists under change of scale, although the values of the coefficients or coupling constants may change. For ϕ^4 it occurs if $d \leq 4$, with $d = 4$ the most interesting case.

1.1.10 The BPHZ Theorem

The BPHZ theorem is both a brilliant historic piece of mathematical physics which gives precise mathematical meaning to the notion of renormalizability, using the mathematics of formal power series, but it is also ultimately a dead end and a bad way to understand and express renormalization. Let us try to explain both statements.

For the massive Euclidean ϕ_4^4 theory we could for instance state the following normalization conditions on the connected functions in momentum space at zero momenta:

$$C^4(0,0,0,0) = -\lambda_{ren}, \tag{1.45}$$

$$C^2(p^2=0) = \frac{1}{m^2_{ren}}, \tag{1.46}$$

$$\frac{d}{dp^2}C^2|_{p^2=0} = -\frac{a_{ren}}{m^4_{ren}}. \tag{1.47}$$

Usually one puts $a_{ren} = 1$ by rescaling the field ϕ.

Using the inversion theorem on formal power series for any *fixed ultraviolet cutoff* κ it is possible to rewrite any formal power series in λ_{bare} with bare propagators $1/(a_{bare}p^2 + m^2_{bare})$ for any Schwinger functions as a formal power series in λ_{ren} with renormalized propagators $1/(a_{ren}p^2 + m^2_{ren})$. The BPHZ theorem then states that that formal perturbative formal power series has finite coefficients order by order when the ultraviolet cutoff κ is lifted. The first proof by Hepp [23] relied on the inductive Bogoliubov's recursion scheme [24]. Then a completely explicit expression for the coefficients of the renormalized series was written by Zimmermann and many followers [25]. The coefficients of that renormalized series can ne written as sums of renormalized Feynman amplitudes. They are similar

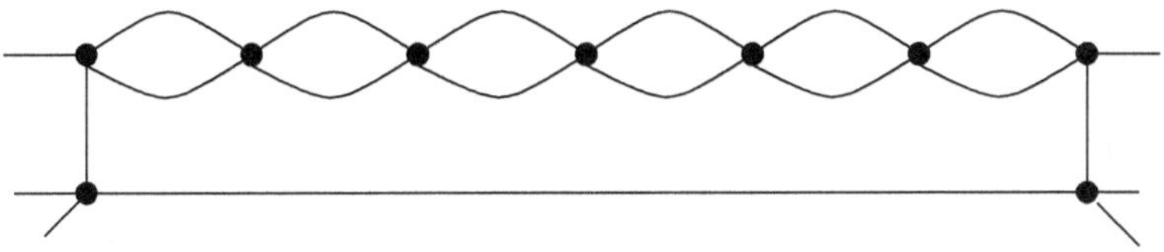

Fig. 1.3 A family of graphs P_n producing a renormalon

to Feynman integrals but with additional subtractions indexed by Zimmermann's forests. Returning to an inductive rather than explicit scheme, Polchinski remarked that it is possible to also deduce the BPHZ theorem from a renormalization group equation and inductive bounds which does not decompose each order of perturbation theory into Feynman graphs [26]. This method was clarified and applied by C. Kopper and coworkers, see [27].

The solution of the difficult "overlapping" divergence problem through Bogoliubov's or Polchinski's recursions and Zimmermann's forests becomes particularly clear in the parametric representation using Hepp's sectors. A Hepp sector is simply a complete ordering of the α parameters for all the lines of the graph. In each sector there is a different classification of forests into packets so that each packet gives a finite integral [28–30].

But from the physical point of view we cannot conceal the fact that purely perturbative renormalization theory is not very satisfying. At least two facts hint at a better theory which lies behind:

- The forest formula seems unnecessarily complicated, with too many terms. For instance in any given Hepp sector only one particular packet of forests is really necessary to make the renormalized amplitude finite, the one which corresponds to the quasi-local divergent subgraphs of *that* sector. The other packets seem useless, a little bit like "junk DNA". They are there just because they are necessary for other sectors. This does not look optimal.
- The theory makes renormalized amplitudes finite, but at tremendous cost! The size of some of these renormalized amplitudes becomes unreasonably large as the size of the graph increases. This phenomenon is called the "renormalon problem". For instance it is easy to check that the renormalized amplitude (at 0 external momenta) of the graphs P_n with six external legs and $n+2$ internal vertices in Fig. 1.3 becomes as large as $c^n n!$ when $n \to \infty$. Indeed at large q the renormalized amplitude $A^R_{G_2}$ in Fig. 1.5 grows like $\log|q|$. Therefore the chain of n such graphs in Fig. 1.3 behaves as $[\log|q|]^n$, and the total amplitude of P_n behaves as

$$\int [\log|q|]^n \frac{d^4q}{[q^2+m^2]^3} \simeq_{n\to\infty} c^n n!\,, \tag{1.48}$$

so that after renormalization some families of graphs acquire so large values that they cannot be resummed! Physically this is just as bad as if infinities were still there.

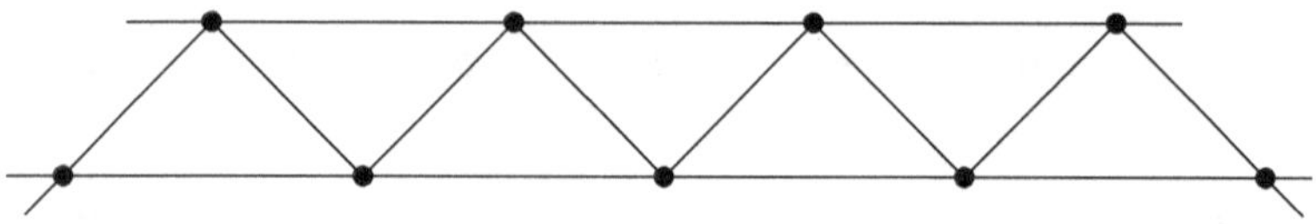

Fig. 1.4 A family of convergent graphs Q_n, that do not produce any renormalon

These two hints are in fact linked. As their name indicates, renormalons are due to renormalization. Families of completely convergent graphs such as the graphs Q_n of Fig. 1.4, are bounded by c^n, and produce no renormalons.

Studying more carefully renormalization in the α parametric representation one can check that renormalons are solely due to the forests packets that we compared to "junk DNA". Renormalons are due to subtractions that are not necessary to ensure convergence, just like the strange $\log |q|$ growth of $A^R_{G_0}$ at large q is solely due to the counterterm in the region where this counterterm is not necessary to make the amplitude finite.

We can therefore conclude that subtractions are not organized in an optimal way by the Bogoliubov recursion. What is wrong from a physical point of view in the BPHZ theorem is to use the size of the graph as the relevant parameter to organize Bogoliubov's induction. It is rather the size of the line momenta that should be used to better organize the renormalization subtractions.

This leads to the point of view advocated in [9]: neither the bare nor the renormalized series are optimal. Perturbation should be organized as a power series in an infinite set of effective expansions, which are related through the RG flow equation. In the end exactly the same contributions are resummed than in the bare or in the renormalized series, but they are regrouped in a much better way.

1.1.11 The Landau Ghost and Asymptotic Freedom

In the case of ϕ^4_4 only the flow of the coupling constants really matters, because the flow of m and of a for different reasons are not very important in the ultraviolet limit:

- The flow of m is governed at leading order by the tadpole. The bare mass m_i^2 corresponding to a finite positive physical mass m_{ren}^2 is negative and grows as λM^{2i} with the slice index i. But since p^2 in the ith slice is also of order M^{2i} but without the λ, as long as the coupling λ remains small it remains much larger than m_i^2. Hence the mass term plays no significant role in the higher slices. It was remarked in [9] that because there are no overlapping problem associated to 1PI two point subgraphs, there is in fact no inconvenience to use the full renormalized m_{ren} all the way from the bare to renormalized scales, with subtractions on 1PI two point subgraphs independent of their scale.

Fig. 1.5 The ϕ^4 connected graphs with $n = 2$, $N = 4$

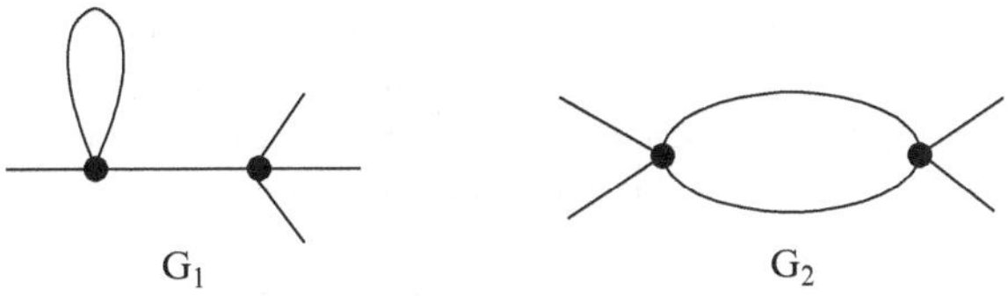

- The flow of a is also not very important. Indeed it really starts at two loops because the tadpole is exactly local. So this flow is in fact bounded, and generates no renormalons. In fact as again remarked in [9] for theories of the ϕ_4^4 type one might as well use the bare value a_{bare} all the way from bare to renormalized scales and perform no second Taylor subtraction on any 1PI two point subgraphs.

But the physics of ϕ_4^4 in the ultraviolet limit really depends of the flow of λ. By a simple second order computation there are only two connected graphs with $n = 2$ and $N = 4$ pictured in Fig. 1.5. They govern at leading order the flow of the coupling constant.

In the commutative ϕ_4^4 theory the graph G_1 does not contribute to the coupling constant flow. This can be seen in many ways, for instance after mass renormalization the graph G_1 vanishes exactly because it contains a tadpole which is not quasi-local but *exactly* local. One can also remark that the graph is one particle reducible. In ordinary translation-invariant, hence momentum-conserving theories, one-particle-reducible quasi-local graphs never contribute significantly to RG flows. Indeed they become very small when the gap i between internal and external scales grows. This is because by momentum conservation the momentum of any one-particle-reducible line ℓ has to be the sum of a finite set of external momenta on one of its sides. But a finite sum of small momenta remains small and this clashes directly with the fact that ℓ being internal its momentum should grow as the gap i grows. Remark that this is no longer be true in non commutative vulcanized $\phi_4^{\star 4}$, because that theory is not translation invariant, and that's why it will ultimately escape the Landau ghost curse.

So in ϕ_4^4 the flow is intimately linked to the sign of the graph G_2 of Fig. 1.5. More precisely, we find that at second order the relation between λ_i and λ_{i-1} is

$$\lambda_{i-1} \simeq \lambda_i - \beta\lambda_i^2 \tag{1.49}$$

(remember the minus sign in the exponential of the action), where β is a constant, namely the asymptotic value of $\sum_{j,j'/\inf(j,j')=i} \int d^4y C_j(x,y) C_{j'}(x,y)$ when $i \to \infty$. Clearly this constant is positive. So for the normal stable ϕ_4^4 theory, the relation (1.49) inverts into

$$\lambda_i \simeq \lambda_{i-1} + \beta\lambda_{i-1}^2, \tag{1.50}$$

so that fixing the renormalized coupling seems to lead at finite i to a large, diverging bare coupling, incompatible with perturbation theory. This is the Landau ghost problem, which affects both the ϕ_4^4 theory and electrodynamics. Equivalently if one keeps λ_i finite as i gets large, $\lambda_0 = \lambda_{ren}$ tends to zero and the final effective theory is "trivial" which means it is a free theory without interaction, in contradiction with the physical observation e.g. of a coupling constant of about $1/137$ in electrodynamics.

But in non-Abelian gauge theories an extra minus sign is created by the algebra of the Lie brackets. This surprising discovery has deep consequences. The flow relation becomes approximately

$$\lambda_i \simeq \lambda_{i-1} - \beta\lambda_i\lambda_{i-1}, \tag{1.51}$$

with $\beta > 0$, or, dividing by $\lambda_i\lambda_{i-1}$,

$$1/\lambda_i \simeq 1/\lambda_{i-1} + \beta, \tag{1.52}$$

with solution $\lambda_i \simeq \frac{\lambda_0}{1+\lambda_0\beta i}$. A more precise computation to third order in fact leads to

$$\lambda_i \simeq \frac{\lambda_0}{1+\lambda_0(\beta i + \gamma \log i + O(1))}. \tag{1.53}$$

Such a theory is called asymptotically free (in the ultraviolet limit) because the effective coupling tends to 0 with the cutoff for a finite fixed small renormalized coupling. Physically the interaction is turned off at small distances. This theory is in agreement with scattering experiments which see a collection of almost free particles (quarks and gluons) inside the hadrons at very high energy. This was the main initial argument to adopt quantum chromodynamics, a non-Abelian gauge theory with $SU(3)$ gauge group, as the theory of strong interactions.

Remark that in such asymptotically free theories which form the backbone of today's standard model, the running coupling constants remain bounded between far ultraviolet "bare" scales and the lower energy scale where renormalized couplings are measured. Ironically the point of view on early renormalization theory as a trick to hide the ultraviolet divergences of QFT into infinite unobservable bare parameters could not turn out to be more wrong than in the standard model. Indeed the bare coupling constants tend to 0 with the ultraviolet cutoff, and what can be farther from infinity than 0?

Recently it has been shown to all orders of perturbation theory that there should be no Landau ghost but an asymptotically safe fixed point for the similar RG flow of the non-commutative Grosse–Wulkenhaar $\phi_4^{\star 4}$ model [31–34]. Therefore this model is a kind of "Ising model" for just renormalizable QFT, that is a simple model in which one can presumably fully mathematically control at last the phenomenon of renormalization in ll its aspects.

1.1.12 Grassmann Representations of Determinants and Pfaffians

Independent Grassmann variables $\chi_1, \ldots, \chi_n$ satisfy complete anticommutation relations

$$\chi_i\chi_j = -\chi_j\chi_i \quad \forall i, j \tag{1.54}$$

so that any function of these variables is a polynomial with highest degree one in each variable.

The rules of Grassmann integration are defined by linearity and

$$\int d\chi_i = 0, \qquad \int \chi_i d\chi_i = 1.$$

plus the rule that all $d\chi$ symbols also anti-commute between themselves and with all χ variables.

The main important facts are

- Any function of Grassmann variables is a polynomial with highest degree one in each variable.
- Pfaffians and determinants can be nicely written as Grassmann integrals.

The determinant of any n by n matrix can indeed be expressed as a Grassmann Gaussian integral over $2n$ independent Grassmann variables which it is convenient to name as $\bar{\psi}_1, \ldots, \bar{\psi}_n, \psi_1, \ldots, \psi_n$, although the bars have nothing yet at this stage to do with complex conjugation. The formula is

$$\det M = \int \prod d\bar{\psi}_i d\psi_i e^{-\sum_{ij} \bar{\psi}_i M_{ij} \psi_j}. \tag{1.55}$$

Remember that for ordinary commuting variables and a positive n by n Hermitian matrix M

$$\frac{1}{\pi^n} \int_{-\infty}^{+\infty} \prod_i d\bar{\phi}_i d\phi_i e^{-\sum_{ij} \bar{\phi}_i M_{ij} \phi_j} = \frac{1}{\det M}. \tag{1.56}$$

In short Grassmann Gaussian measures are simpler than ordinary Gaussian measures for two main reasons:

- Grassmann Gaussian measures are associated to any matrix M, there is no positivity requirement for M like for ordinary Gaussian measures.
- Their normalization directly computes the determinant of M, not the inverse (square-root of) the determinant of M. This is essential in many areas where factoring out this determinant is desirable; it explains in particular the success of Grassmann and supersymmetric functional integrals in the study of *disordered systems*.

The stubborn reader which remembers the square-root formula in (1.1) and would like to understand the corresponding "*real* version" of (1.55) is rewarded by the beautiful theory of Pfaffians. Clearly commuting Gaussian real integrals involve symmetric matrices, but Grassmann Gaussian with only n "real" integrals must involve n by n antisymmetric matrices.

The Pfaffian $\mathrm{Pf}(A)$ of an *antisymmetric* matrix A is defined by

$$\det A = [\mathrm{Pf}(A)]^2. \tag{1.57}$$

and is known to be polynomila in the coefficients of A. This fact is recovered easily by writing it as

$$\mathrm{Pf}(A) = \int d\chi_1 \ldots d\chi_n e^{-\sum_{i<j} \chi_i A_{ij} \chi_j} = \int d\chi_1 \ldots d\chi_n e^{-\frac{1}{2}\sum_{i,j} \chi_i A_{ij} \chi_j}. \quad (1.58)$$

Indeed we have

$$\det A = \int \prod_i d\bar{\psi}_i d\psi_i e^{-\sum_{ij} \bar{\psi}_i A_{ij} \psi_j}. \quad (1.59)$$

Performing the change of variables (which a posteriori justifies the complex notation)

$$\bar{\psi}_i = \frac{1}{\sqrt{2}}(\chi_i - i\omega_i), \quad \psi_i = \frac{1}{\sqrt{2}}(\chi_i + i\omega_i), \quad (1.60)$$

whose Jacobian is i^{-n}, the new variables χ and ω are again independent Grassmann variables. Now a short computation using $A_{ij} = -A_{ji}$ gives

$$\det A = i^{-n} \int \prod_i d\chi_i d\omega_i e^{-\sum_{i<j} \chi_i A_{ij} \chi_j - \sum_{i<j} \omega_i A_{ij} \omega_j}$$

$$= \int \prod_i d\chi_i e^{-\sum_{i<j} \chi_i A_{ij} \chi_j} \prod_i d\omega_i e^{-\sum_{i<j} \omega_i A_{ij} \omega_j}, \quad (1.61)$$

where we used that $n = 2p$ has to be even and that a factor $(-1)^p$ is generated when changing $\prod_i d\chi_i d\omega_i$ into $\prod_i d\chi_i \prod_i d\omega_i$. Equation (1.61) shows why $\det A$ is a perfect square and proves (1.58).

A useful Lemma is:

Lemma 1.1. *The determinant of a matrix $D + A$ where D is diagonal and A antisymmetric has a "quasi-Pfaffian" representation*

$$\det(D + A) = \int \prod_i d\chi_i d\omega_i e^{-\sum_i \chi_i D_{ii} \omega_i - \sum_{i<j} \chi_i A_{ij} \chi_j + \sum_{i<j} \omega_i A_{ij} \omega_j}. \quad (1.62)$$

Proof. The proof consists in performing the change of variables (1.60) and canceling carefully the i factors.

There are also *normalized* Grassmann Gaussian integrals which may be expressed formally as

$$d\mu_M = \frac{\prod d\bar{\psi}_i d\psi_i e^{-\sum_{ij} \bar{\psi}_i M_{ij}^{-1} \psi_j}}{\int \prod d\bar{\psi}_i d\psi_i e^{-\sum_{ij} \bar{\psi}_i M_{ij}^{-1} \psi_j}}. \quad (1.63)$$

and again are characterized by their two point function or covariance

$$\int \bar{\psi}_i \psi_j d\mu_M = M_{ij}. \tag{1.64}$$

plus the Grassmann–Wick rule that n-point functions are expressed as sum over Wick contractions with *signs*. □

For a much more detailed introduction to the rules of Grassmann calculus in QFT, we refer to [35].

1.1.13 Trees, Forests and the Parametric Representation

Classical evolution can be expanded perturbatively into sums indexed by trees whether in quantum field theory the loops of Feynman graphs are essential.

The hidden trees of the classical system inside QFT can be revealed only under scale analysis, since they do *not* correspond to ordinary spanning trees of the graphs, but to the abstract inclusion relations of short range effects (high energy quasi local subgraphs) inside larger ones. This point of view has been progressively formalized over the years from Bogoliubov to Zimmermann to the most recent formalization by D. Kreimer and A. Connes in terms of Hopf algebras.

But *ordinary spanning trees* of a connected graph also enter in a fascinating way in the computation of its amplitude. Since the heat kernel is quadratic it is possible to explicitly compute all spatial integrations in a Feynman amplitude. One obtains the so-called parametric representation. The result is expressed in terms of topological or so-called "Symanzik" polynomials [36, 37].

The amplitude of an amputated graph G with external momenta p is, up to a normalization, in space-time dimension D:

$$A_G(p) = \delta(\sum p) \int_0^\infty \frac{e^{-V_G(p,\alpha)/U_G(\alpha)}}{U_G(\alpha)^{D/2}} \prod_l (e^{-m^2 \alpha_l} d\alpha_l) . \tag{1.65}$$

The first and second Symanzik polynomials U_G and V_G are

$$U_G = \sum_T \prod_{l \notin T} \alpha_l , \tag{1.66a}$$

$$V_G = \sum_{T_2} \prod_{l \notin T_2} \alpha_l (\sum_{i \in E(T_2)} p_i)^2 , \tag{1.66b}$$

where the first sum is over spanning trees T of G and the second sum is over two trees T_2, i.e. forests separating the graph in exactly two connected components

$E(T_2)$ and $F(T_2)$; the corresponding Euclidean invariant $(\sum_{i\in E(T_2)} p_i)^2$ is, by momentum conservation, also equal to $(\sum_{i\in F(T_2)} p_i)^2$.

The proof of relations (1.66a)–(1.66b) is a special case of the Tree matrix Theorem, which we now explain following [38]

Theorem 1.1 (Tree Matrix Theorem). *Let A be an n by n matrix such that*

$$\sum_{i=1}^{n} A_{ij} = 0 \ \forall j \ . \tag{1.67}$$

Obviously $\det A = 0$. *But let A^{11} be the matrix A with line 1 and column 1 deleted. Then*

$$\det A^{11} = \sum_{T} \prod_{\ell\in T} A_{i_\ell, j_\ell}, \tag{1.68}$$

where the sum runs over all directed trees on $\{1,\dots,n\}$, directed away from root 1.

This theorem has both a positivity and a democracy aspect: all trees contribute with positive, equal weights to the determinant.

Proof of Theorem 1.1: We use Grassmann variables to write the determinant of a matrix with one line and one raw deleted as a Grassmann integral with two corresponding sources:

$$\det A^{11} = \int (\mathrm{d}\overline{\psi}\mathrm{d}\psi)\ (\psi_1\overline{\psi}_1)e^{-\overline{\psi}A\psi}. \tag{1.69}$$

The trick is to use (1.67) to write

$$\overline{\psi}A\psi = \sum_{i,j=1}^{n} (\overline{\psi}_i - \overline{\psi}_j)A_{ij}\psi_j \tag{1.70}$$

and to obtain

$$\det A^{11} = \int \mathrm{d}\overline{\psi}\mathrm{d}\psi\ (\psi_1\overline{\psi}_1)\exp\left(-\sum_{i,j=1}^{n} A_{ij}(\overline{\psi}_i - \overline{\psi}_j)\psi_j\right) \tag{1.71}$$

$$= \int \mathrm{d}\overline{\psi}\mathrm{d}\psi\ (\psi_1\overline{\psi}_1)\left[\prod_{i,j=1}^{n} \left(1 - A_{ij}(\overline{\psi}_i - \overline{\psi}_j)\psi_j\right)\right] \tag{1.72}$$

by the Pauli exclusion principle. We now expand to get

$$\det A^{11} = \sum_{\mathcal{G}} \left(\prod_{\ell=(i,j)\in\mathcal{G}} (-A_{ij})\right) \Omega_{\mathcal{G}} \tag{1.73}$$

where $\mathcal{G}$ is *any* subset of $[n] \times [n]$, and we used the notation

$$\Omega_{\mathcal{G}} \stackrel{\text{def}}{=} \int \mathrm{d}\overline{\psi}\mathrm{d}\psi \; (\psi_1 \overline{\psi}_1) \left(\prod_{(i,j) \in \mathcal{G}} \left[(\overline{\psi}_i - \overline{\psi}_j) \psi_j \right] \right). \tag{1.74}$$

The tree matrix theorem then follows from the following

Lemma 1.2. $\Omega_{\mathcal{G}} = 0$ *unless the graph* $\mathcal{G}$ *is a tree directed away from 1 in which case* $\Omega_{\mathcal{G}} = 1$.

Proof of the Lemma. Trivially, if (i,i) belongs to $\mathcal{G}$, then the integrand of $\Omega_{\mathcal{G}}$ contains a factor $\overline{\psi}_i - \overline{\psi}_i = 0$ and therefore $\Omega_{\mathcal{G}}$ vanishes.

But the crucial observation is that if there is a loop in $\mathcal{G}$ then again $\Omega_{\mathcal{G}} = 0$. This is because then the integrand of $\Omega_{\mathcal{F},\mathcal{R}}$ contains the factor

$$\overline{\psi}_{\tau(k)} - \overline{\psi}_{\tau(1)} = (\overline{\psi}_{\tau(k)} - \overline{\psi}_{\tau(k-1)}) + \cdots + (\overline{\psi}_{\tau(2)} - \overline{\psi}_{\tau(1)}). \tag{1.75}$$

Now, upon inserting this telescoping expansion of the factor $\overline{\psi}_{\tau(k)} - \overline{\psi}_{\tau(1)}$ into the integrand of $\Omega_{\mathcal{F},\mathcal{R}}$, the latter breaks into a sum of $(k-1)$ products. For each of these products, there exists an $\alpha \in \mathbb{Z}/k\mathbb{Z}$ such that the factor $(\overline{\psi}_{\tau(\alpha)} - \overline{\psi}_{\tau(\alpha-1)})$ appears *twice*: once with the $+$ sign from the telescopic expansion of $(\overline{\psi}_{\tau(k)} - \overline{\psi}_{\tau(1)})$, and once more with a $+$ (resp. $-$) sign if $(\tau(\alpha), \tau(\alpha-1))$ (resp. $(\tau(\alpha-1), \tau(\alpha))$) belongs to $\mathcal{F}$. Again, the Pauli exclusion principle entails that $\Omega_{\mathcal{G}} = 0$.

Now every connected component of $\mathcal{G}$ must contain 1, otherwise there is no way to saturate the $d\psi_1$ integration.

This means that $\mathcal{G}$ has to be a directed tree on $\{1, \ldots n\}$. It remains only to see that $\mathcal{G}$ has to be directed away from 1, which is not too difficult. □

Relations (1.66a)–(1.66b) follow rather easily from the tree matrix theorem and the *direct representation of Feynman amplitudes (1.77).*

In [21] a deeper proof of these relations is given. It relies on the more canonical *phase-space parametric representation*, which we briefly describe now. Let us limit ourselves to "semi-regular" graphs, which have no "tadpoles" that it no line starting and ending at the same vertex. These graphs (once their lines have been *oriented* in an arbitrary way) are nicely characterized by their incidence matrix, which is a regular $l(G)$ by $n(G)$ matrix $\epsilon_{\ell v}$ with

$$\begin{aligned} \epsilon_{\ell v} &= -1 \text{ if line } \ell \text{ exits vertex } v, \\ \epsilon_{\ell v} &= +1 \text{ if line } \ell \text{ enters vertex } v, \\ \epsilon_{\ell v} &= 0 \quad \text{otherwise} . \end{aligned} \tag{1.76}$$

There are also external momenta p_f, $f = 1, \cdots, N$, which we could also also orient through a matrix ϵ_{fv}.

The momentum parametric representation then writes

$$A^T_G(p_1,\ldots,p_N) = \delta(\sum_{f,v} \epsilon_{fv} p_f) \int \prod_{\ell=1}^{l(G)} d\alpha_\ell d^d k_\ell e^{-\alpha_\ell (k_\ell^2 + m^2)}$$
$$\times \prod_{v=1}^{n(G)-1} \delta(\epsilon_{fv} p_f + \epsilon_{\ell v} k_\ell).$$

But there is no canonical way to solve for the delta functions, something known in physics as the procedure of *momentum attribution*. So it is better to rewrite these amplitudes in the *phase-space parametric representation*

$$A^T_G(p_1,\ldots,p_N) = \int \prod_{\ell=1}^{l(G)} [d\alpha_\ell e^{-\alpha_\ell m^2} d^d k_\ell] \prod_{v=1}^{V-1} d^d x_v e^{-\alpha_\ell k_\ell^2 + 2i(p_f \epsilon_{vf} x_v + k_\ell \epsilon_{v\ell} x_v)},$$

Integrating over momenta leads to the *direct space parametric representation*:

$$A^T_G(p_1,\ldots,p_N) = \int \prod_{\ell=1}^{l(G)} d\alpha_\ell \frac{e^{-\alpha_\ell m^2}}{\alpha_\ell^{d/2}} \prod_{v=1}^{n(G)-1} d^d x_v e^{2ip_f \epsilon_{vf} x_v - x_v \cdot x_{v'} \epsilon_{\ell v} \epsilon_{\ell v'}/\alpha_\ell}.$$

In [21] it is shown how the representation (1.77) together with the quasi-Pfaffian representation of Lemma 1.1 leads to deletion-contraction relations for the Symanzik polynomials which allow to compute (1.66a)–(1.66b) from the theory of the universal Tutte polynomial.

1.1.14 BKAR Forest Formula

Since we want to implement Renormalization Group in a non-perturbative or constructive way, we need tools to compute connected functions in a non-perturbative way, with the right scaling properties for the convergence radius of the expansion. For instance in the just renormalizable case, we need a convergence radius in the coupling constant which is *uniform in the scale index.*

The main such tool is a canonical *forest formula* [39, 40] which allows to package a perturbative expansion in terms of trees rather than Feynman graphs. The advantage was already mentioned several times: the species of trees does not *proliferate* [19, 20] at large orders, in contrast with the species of Feynman graphs.

Consider n points; the set of pairs P_n of such points which has $n(n-1)/2$ elements $\ell = (i,j)$ for $1 \le i < j \le n$, and a smooth function f of $n(n-1)/2$ variables x_ℓ, $\ell \in \mathcal{P}_n$. Noting ∂_ℓ for $\frac{\partial}{\partial x_\ell}$, the standard forest formula is

$$f(1,\ldots,1) = \sum_{\mathscr{F}} [\prod_{\ell \in \mathscr{F}} \int_0^1 dw_\ell]([\prod_{\ell \in \mathscr{F}} \partial_\ell] f)[x_\ell^{\mathscr{F}}(\{w_{\ell'}\})] \tag{1.77}$$

where

- The sum over $\mathscr{F}$ is over forests over the n vertices, including the empty one.
- $x_\ell^{\mathscr{F}}(\{w_{\ell'}\})$ is the infimum of the $w_{\ell'}$ for ℓ' in the unique path from i to j in $\mathscr{F}$, where $\ell = (i,j)$. If there is no such path, $x_\ell^{\mathscr{F}}(\{w_{\ell'}\}) = 0$ by definition.
- The symmetric n by n matrix $X^{\mathscr{F}}(\{w\})$ defined by $X_{ii}^{\mathscr{F}} = 1$ and $X_{ij}^{\mathscr{F}} = x_{ij}^{\mathscr{F}}(\{w_{\ell'}\})$ for $1 \le i < j \le n$ is positive.

This formula can be viewed as a tool to associate to any pair made of a graph G and a spanning forest $F \subset G$ a unique rational number or weight $w(G,F)$ between 0 and 1, called the relative weight of T in G. These weights are barycentric or percentage factors, ie for any G

$$\sum_{F \subset G} w(G,F) = 1. \tag{1.78}$$

The numbers $w(G,F)$ are multiplicative over disjoint unions.[6] Hence it is enough to give the formula for (G,F) only when G is *connected* and $F = T$ is a spanning tree in it.[7] The definition of these weights is

Definition 1.1.

$$w(G,T) = \prod_{\ell \in T} \int_0^1 \prod_{\ell \in T} dw_\ell \prod_{\ell \notin T} x_\ell^T(\{w\}) \tag{1.79}$$

where $x_\ell^T(\{w\})$ is again the infimum over the $w_{\ell'}$ parameters over the lines ℓ' forming the *unique* path in T joining the ends of ℓ.

Consider the expansion in terms of Feynman amplitudes of a connected quantity S. The most naive way to reorder Feynman perturbation theory according to trees rather than graphs is to insert for each graph the relation (1.78)

$$S = \sum_G A_G = \sum_G \sum_{T \subset G} w(G,T)\mathscr{A}_G \tag{1.80}$$

and exchange the order of the sums over S and T. Hence it writes

$$S = \sum_T \mathscr{A}_T, \quad \mathscr{A}_T = \sum_{G \supset T} w(G,T)\mathscr{A}_G. \tag{1.81}$$

[6]And also over vertex joints of graphs, just as in the universality theorem for the Tutte polynomial.

[7]It is enough in fact to compute such weights for one-particle irreducible and one-vertex-irreducible graphs, then multiply them in the appropriate way for the general case.

This rearranges the Feynman expansion according to trees, but each tree has the same number of vertices as the initial graph. Hence it reshuffles the various terms of a *given, fixed* order of perturbation theory. Remark that if the initial graphs have say degree 4 at each vertex, only trees with degree less than or equal to 4 occur in the rearranged tree expansion.

For Fermionic theories this is typically sufficient and one has for small enough coupling

$$\sum_T |\mathscr{A}_T| < \infty \tag{1.82}$$

because Fermionic graphs essentially mostly compensate each other at a fixed order by Pauli's principle; mathematically this is because these graphs form a determinant and the size of a determinant is much less than what its permutation expansion usually suggests. This is well known [41–43].

But this repacking fails completely for Bosonic theories, because the only compensations there occur between graphs of different orders. Hence if we perform this naive reshuffling, eg on the ϕ_0^4 theory we would still have

$$\sum_T |\mathscr{A}_T| = \infty. \tag{1.83}$$

Recently a new expansion called the Loop Vertex Expansion has been found [16] which overcomes this difficulty by exchanging the role of vertices and propagators before applying the forest formula. It can also be seen as a combination of the forest formula with the so-called intermediate field representation, which expands into essentially *square roots* of a stable Bosonic interaction. We refer the reader to [16, 17, 19, 20, 44] but wont review this expansion here, since from now on we are mostly going to deal with Fermions.

1.1.15 Gram and Hadamard Bounds

These two bounds on a determinant are often confused!

The Gram bound applies to a matrix $A = a_{ij}$ whose entries are scalar product. This means we suppose that there exists some Hilbert space H and $2n$ vectors f_i, $i = 1, \cdots, n$, g_j, $j = 1, \cdots, n$ with

$$a_{ij} =< f_i, g_j >_H . \tag{1.84}$$

The Gram bound states

$$|\det A| \leq \prod_{i=1}^n \|f_i\|_H \prod_{j=1}^n \|g_j\|_H . \tag{1.85}$$

Of course *any* matrix A can always be written of the Gram type, eg with $H = \mathbb{R}^n$, $f_i = (a_{i,k})$ and $g_j = \delta_{j,k}$, $k = 1, \cdots, n$, or conversely. Hence there are two corresponding asymmetric Hadamard bounds, one for rows and one for columns:

$$|\det A| \le \prod_{i=1}^{n} \sqrt{\sum_{j=1}^{n} a_{ij}^2}, \tag{1.86}$$

$$|\det A| \le \prod_{j=1}^{n} \sqrt{\sum_{i=1}^{n} a_{ij}^2} \tag{1.87}$$

and also a symmetric Hadamard bound involving the supremum of the matrix elements:

$$|\det A| \le n^{n/2} \big(\sup_{i,j} |a_{ij}|\big)^n. \tag{1.88}$$

Remark that the symmetric Hadamard bound means that a determinant of a large matrix is always *much smaller* than what its permutation expansion plus naive bounds would suggest, which is the "stupid bound"

$$|\det A| \le n! \big(\sup_{i,j} |a_{ij}|\big)^n. \tag{1.89}$$

This difference in constructive theory is essential. Indeed for Fermionic theories with bounded propagators and a quartic interaction, the matrix A at nth order of perturbation is a $2n \times 2n$ matrix, with propagators as matrix elements, and there is a $1/n!$ symmetry factor. Hence the bound (1.89) would lead to believe that the radius of convergence of the partition function is 0, like in the Bosonic case. But the Hadamard bound (1.88) proves that it is at least finite. Moreover usually it is possible to write the propagators as scalar products in $L^2(\mathbb{R})^d$ of functions which also have bounded L^2 norms.[8] In that case the Gram bound (1.85) shows that the partition function is in fact an *entire* function, as it shows no factorial dependence at all as $n \to \infty$!

1.1.16 Single Scale Constructive Theory for a Toy Model

Consider a just-renormalizable QFT theory. The key problem is to compute connected quantities with an expansion which converges for a small coupling constant, with a propagator limited to a single renormalization group scale, uniformly in the slice index. In the simple Fermionic case, this can be done through applying first the BKAR formula then checking convergence through a Gram bound.

[8]This is usually easily done by taking some kind of "square roots" in momentum space.

To discuss the type of Fermionic models met in condensed matter it is appropriate to consider first a toy Fermionic d-dimensional QFT model. It is made of a single an infrared slice with N colors. Suppose the propagator is diagonal in color space and satisfies the bound

$$|C_{j,ab}(x,y)| \leq \delta_{ab} \frac{M^{-dj/2}}{\sqrt{N}} e^{-M^{-j}|x-y|}. \tag{1.90}$$

We say that the interaction is of the vector type (or Gross–Neveu type) if it is of the form

$$V = \lambda \int d^d x (\sum_{a=1}^{N} \bar{\psi}_a(x)\psi_a(x))(\sum_{b=1}^{N} \bar{\psi}_b(x)\psi_b(x)) \tag{1.91}$$

where λ is the coupling constant.

We claim that

Lemma 1.3. *The perturbation theory for the connected functions of this single slice model has a radius of convergence in λ which is uniform in j and N.*

To prove this lemma, expands the partition function $Z(\Lambda)$ through the forest formula, and take the logarithm to obtain a tree formula for the pressure

$$p = \lim_{\Lambda \to \infty} \frac{1}{|\Lambda|} \log Z(\Lambda). \tag{1.92}$$

This is completely straightforward, the only difficulty being notational. Using the notations of [42]

$$p = \lim_{\Lambda \to \infty} \frac{1}{|\Lambda|} (\int d\mu_C(\psi, \bar{\psi}) e^{S_\Lambda(\bar{\psi}_a, \psi_a)}) \tag{1.93}$$

$$= \sum_{n=0}^{\infty} (\lambda^n / N^n n!) \sum_{a_1,\ldots,a_n,b_1,\ldots,b_n=1}^{N} \sum_{\mathcal{T}} \sum_{\Omega} \epsilon(\mathcal{T}, \Omega) (\prod_{l \in \mathcal{T}} \int_0^1 dw_l)$$

$$\int_{\mathbb{R}^{nd}} dx_1 \ldots dx_n \delta(x_1 = 0) \prod_{l \in \mathcal{T}} \big(C(x_{i(l)}, x_{j(l)})\delta_l\big) \times \det_{\text{remaining}} (C_{\alpha\beta})_{\alpha \in A, \beta \in B}.$$

The sum over the a_i's and b_i's are over the colors of the fields and anti-fields of the vertices obtained by expanding the interaction and of the form:

$$\bar{\psi}_{a_i}(x_i)\psi_{a_i}(x_i)\bar{\psi}_{b_i}(x_i)\psi_{b_i}(x_i) \tag{1.94}$$

with $1 \leq i \leq n$. The sum over $\mathcal{T}$ is over all trees which connect together the n vertices at $x_1, \ldots, x_n$. The sum over Ω is over the compatible ways of realizing the bonds $l = \{i, j\} \in \mathcal{T}$ as contractions of a ψ and $\bar{\psi}$ between the vertices i and j (compatible means that we do not contract twice the same field or anti-field).

$\epsilon(\mathscr{T}, \Omega)$ is a sign which is not important for the bound (see [42] for its explicit computation). For any $l \in \mathscr{T}$, $i(l) \in \{1, \ldots, n\}$ labels the vertex where the field, contracted by the procedure Ω concerning the link l, was chosen. Likewise $j(l)$ is the label for the vertex containing the contracted anti-field. δ_l is 1 if the colors (among $a_1, \ldots, a_n, b_1, \ldots, b_n$) of the field and anti-field contracted by l are the same and else is 0. Finally the matrix $(C_{\alpha\beta})_{\alpha,\beta}$ of the remaining "loop lines" is defined in the following manner.

The row indices α label the $2n$ fields produced by the n vertices, so that $\alpha = (i, \sigma)$ with $1 \leq i \leq n$ and σ takes two values 1 or 2 to indicate whether the field is the second or the fourth factor in (1.94) respectively.

The column indices β label in the same way the $2n$ anti-fields, so that $\beta = (j, \tau)$ with $1 \leq j \leq n$ and $\tau = 1$ or 2 according to whether the anti-field is the first or the third factor in (1.94) respectively. The α's and β's are ordered lexicographically. We denote by $c(i, \sigma)$ the color of the field labeled by (i, σ) that is a_i, if $\sigma = 1$, and b_i if $\sigma = 2$. We introduce the similar notation $\bar{c}(j, \tau)$ for the color of an anti-field. Now

$$C_{(i,\sigma)(j,\tau)} = w_{ij}^{\mathscr{T},BK}(\mathbf{w}) C(x_i, x_j) \delta(c(i,\sigma), \bar{c}(j,\tau)). \tag{1.95}$$

Finally each time a field (i, σ) is contracted by Ω the corresponding row is deleted from the $2n \times 2n$ matrix $(C_{\alpha\beta})$. Likewise, for any contracted anti-field the corresponding column is erased. A and B denote respectively the set of remaining rows and the set of remaining columns. The minor determinant featuring as $\det_{remaining}$ in (1.94) is now $\det(b_{\alpha\beta})_{\alpha \in A, \beta \in B}$ which is $(n+1) \times (n+1)$. Indeed for each of the $n - 1$ links of $\mathscr{T}$, a row and a column have been erased.

Suppose we have written

$$C_j(x_k, y_m) < f_{j,k}, g_{j,m} >_{L^2} \tag{1.96}$$

(this is realized through $f_{j,k} = f_j(x_k, \cdot)$ and $g_{j,m} = g_j(\cdot, y_m)$ if $\hat{f}_j(p).\hat{g}_j(p) = \hat{C}_j(p)$). and that the power counting is conserved by taking square-roots, namely that the L^2 norms of g_j and f_j scale as $\frac{M^{-dj/4}}{N^{1/4}}$.

Then applying the Gram inequality

$$|\det[C_{j,ab}()]_{\text{remaining}}| \leq \prod_{\text{anti-fields } k} ||f_{j,k}|| \prod_{\text{fields } m} ||g_{j,m}|| \tag{1.97}$$

will lead to the proof that the radius of convergence of the pressure is uniform in j and N for this toy model.

Indeed the following lemma shows that the presence of the weakening factors w does not change the outcome of the Gram bound.

Lemma. *Let $\mathscr{A} = (a_{\alpha\beta})_{\alpha,\beta}$ be a Gram matrix: $a_{\alpha\beta} = < f_\alpha, g_\beta >$ for some inner product $< .,. >$. Suppose each of the indices α and β is of the form (i, σ) where the first index i, $1 \leq i \leq n$, has the same range as the indices of the positive matrix $(w_{ij}^{\mathscr{F},BK}(\mathbf{w}))_{ij}$, and σ runs through some other index set Σ.*

Let $\mathscr{C} = (C_{\alpha\beta})_{\alpha,\beta}$ be the matrix with entries $C_{(i,\sigma)(j,\tau)} = w_{ij}^{\mathscr{F},BK}(\mathbf{w}). < f_{(i,\sigma)}, g_{(j,\tau)} >$ and let $(C_{\alpha\beta})_{\alpha\in A,\beta\in B}$ be some square matrix extracted from $\mathscr{C}$, then for any $\mathbf{w}$ we have the Gram inequality:

$$|\det(C_{\alpha\beta})_{\alpha\in A,\beta\in B}| \leq \prod_{\alpha\in A} ||f_\alpha|| \prod_{\beta\in B} ||g_\beta||. \tag{1.98}$$

Proof. Indeed we can take the symmetric square root v of the positive matrix $w^{\mathscr{F},BK}$ so that $w_{ij}^{\mathscr{F},BK} = \sum_{k=1}^{n} v_{ik}v_{kj}$. Let us denote the components of the vectors f and g, in an orthonormal basis for the scalar product $< .,. >$ with q elements, by $f^m_{(i,\sigma)}$ and $g^m_{(j,\tau)}$, $1 \leq m \leq q$. (Indeed even if the initial Hilbert space is infinite dimensional, the problem is obviously restricted to the finite dimensional subspace generated by the finite set of vectors f and g). We then define the tensors $F_{(i,\sigma)}$ and $G_{(j,\tau)}$ with components $F^{km}_{(i,\sigma)} = v_{ik} f^m_{(i,\sigma)}$ and $G^{km}_{(j,\tau)} = v_{jk} g^m_{(j,\tau)}$ where $1 \leq k \leq n$ and $1 \leq m \leq q$. Now considering the tensor scalar product $< .,. >_T$ we have

$$< F_{(i,\sigma)}, G_{(j,\tau)} >_T = \sum_{k=1}^{n}\sum_{m=1}^{q} v_{ik}v_{jk} f^m_{(i,\sigma)} g^m_{(j,\tau)} = b_{(i,\sigma)(j,\tau)}. \tag{1.99}$$

By Gram's inequality using the $< .,. >_T$ scalar product we get

$$|\det(C_{\alpha\beta})_{\alpha\in A,\beta\in B})| \leq \prod_{\alpha\in A} ||f_\alpha||_T \prod_{\beta\in B} ||g_\beta||_T. \tag{1.100}$$

Also

$$\begin{aligned} ||F_{(i,\sigma)}||_T^2 &= \sum_{k=1}^{n}\sum_{m=1}^{q} (F^{km}_{(i,\sigma)})^2 = \sum_{k=1}^{n}\sum_{m=1}^{q} v_{ik}^2 (f^m_{(i,\sigma)})^2 \\ &= w_{ii} \sum_{m=1}^{q} (f^m_{(i,\sigma)})^2 = ||f_{(i,\sigma)}||^2, \end{aligned} \tag{1.101}$$

since $w_{ii} = 1$ for any i, $1 \leq i \leq n$. □

Let us apply the Gram bound and this last Lemma to bound the determinant in (1.94).

- There is a factor $M^{-dj/2}$ per line, or $M^{-dj/4}$ per field ie entry of the loop determinant. This gives a factor M^{-dj} per vertex.
- There is a factor M^{+dj} per vertex spatial integration (minus one).

Hence the λ radius of convergence is uniform in j.

- There is a factor $N^{-1/2}$ per line, or $N^{-1/4}$ per field ie entry of the loop determinant. This gives a factor N^{-1} per vertex.
- There is a factor N per vertex (plus an extra one).

This leads to a bound in $N.M^{2j}[c\lambda]^n$ for the nth order of the pressure, hence to a radius of convergence at least $1/c$. As expected, this is a bound uniform in the slice index j. Hence the λ radius of convergence is uniform in N.

The last item, namely the factor N^{n+1} for the colors sums is the only one not obvious to prove. Indeed don't know all the graph, but only a spanning tree. We need to organize the sum over the colors *from the leaves to the root of the tree*. In this way the pay a factor N at each leaf to know the color index which *does not go towards the root*, then prune the leaf and iterate. The last vertex (the root) is the only special one as it costs *two* N factors.

Let us remark that to treat the corresponding toy model in the Bosonic case the standard constructive method would be to perform a cluster expansion with respect to a lattice of cubes, then a Mayer expansion which further removed the remaining hardcore constraints with respect to the cubes [9]. Both expansions needed to use the forest formula. This was simplified by the invention of the Loop vertex expansion, in which cubes, cluster and Mayer expansions are no longer needed. In addition the Loop vertex model leads to uniform bounds also for *matrix* toy models, a result which cannot be obtained up to now with other methods [16].

1.2 Interacting Fermions in Two Dimensions

1.2.1 Introduction

One of the main achievements in renormalization theory has been the extension of the renormalization group of Wilson (which analyzes long-range behavior governed by simple scaling around the point singularity $p = 0$ in momentum space) to long-range behavior governed by extended singularities [45–47]. This very natural and general idea is susceptible of many applications in various domains, including condensed matter (reviewed here, in which the extended singularity is the Fermi surface) but also other ones such as diffusion in Minkowski space (in which the extended singularity is the mass shell). In this section we will discuss interacting Fermions models such as those describing the conduction electrons in a metal.

The key features which differentiate electrons in condensed matter from Euclidean field theory, and make the subject in a way mathematically richer, is that space-time rotation invariance is broken, and that particle density is finite. This finite density of particles creates the Fermi sea: particles fill states up to an energy level called the Fermi surface.

The field theory formalism is the best tool to isolate fundamental issues such as the existence of non-perturbative effects. In this formalism the usual Hamiltonian point of view with operators creating electrons or holes is superseded by the more synthetic point of view of anti-commuting Fermionic fields with two spin indices and arguments in $d + 1$ dimensional space-time. Beware however of the QFT-convention to always call dimension the dimension of *space-time*, whether from now on we have to stick to the usual condensed matter convention which is to always call dimension the dimension of *space only*. So one dimensional interacting Fermions correspond at zero temperature to a two dimensional QFT, two dimensional Fermions correspond at zero temperature to a three dimensional QFT and so on.

After the discovery of high-T_c superconductivity, a key question emerged. Do interacting Fermions in two dimensions (above their low-temperature phase) resemble more three dimensional Fermions, i.e. the Fermi liquid, or one dimensional Fermions, i.e. the Luttinger liquid? The short answer to this controversial question is that it was solved rigorously by mathematical physics and that the answer depends on the shape of the Fermi surface. Interacting Fermions with a round Fermi surface behave more like three dimensional Fermi liquids, whether interacting Fermions with the square Fermi surface of the Hubbard model at half-filling behave more like a one-dimensional Luttinger liquid.

This statement has been now proved in full mathematical rigor, beyond perturbation theory, in the series of works [48–54].

See also the interesting case of the honeycomb lattice at half-filling (graphene) for 2D Luttinger liquid behavior [55, 56].

The existence of usual 2D interacting Fermi liquids was established in [48, 49] using the mathematically precise criterion of Salmhofer [57] in the case of a temperature infra-red regulator. Using a magnetic field regulator that breaks parity invariance of the Fermi surface it was also established in the initial sense of a discontinuity at the Fermi surface in the series of papers [58].

1.2.2 The Models: J_2, J_3, H_2...

We consider a gas of Fermions in thermal equilibrium at temperature T, with coupling constant λ. The free propagator for this model is, for continuum models

$$\hat{C}_{ab}(k) = \delta_{ab} \frac{\eta(k)}{ik_0 - e(\mathbf{k})} \tag{1.102}$$

with $\mathbf{k}$ being the d-dimensional momentum, $e(\mathbf{k}) = \epsilon(\mathbf{k}) - \mu$, $\epsilon(\mathbf{k})$ being the kinetic energy and μ the chemical potential. $\eta(k)$ is a fixed ultraviolet cutoff which smoothly cuts out large momenta. There are also similar models on lattices where the ultraviolet cutoff is provided by the lattice. The $a, b \in \{\uparrow, \downarrow\}$ index is for spin hence can take two values (remember spin is treated non-relativistically).

At finite temperature, Fermionic fields have to satisfy anti-periodic boundary conditions. Hence the component k_0 in (1.102) can take only discrete values (called the Matsubara frequencies): so the integral over k_0 is really a discrete sum over n.

These Matsubara frequencies are:

$$k_0 = \frac{2n+1}{\beta\hbar}\pi, \quad n \in \mathbb{Z} \tag{1.103}$$

where $\beta = (kT)^{-1}$. For any n we have $k_0 \neq 0$, so that the denominator in $C(k)$ can never be 0. This is why the temperature provides a natural infrared cut-off.

We can think of k_0 as the Fourier dual to an imaginary Euclidean-time continuous variable taking values in a circle, with length proportional to inverse temperature β. When $T \to 0^+$, (which means $\beta \to +\infty$), k_0 becomes a continuous variable, the corresponding discrete sum becomes an integral, and the corresponding propagator $C_0(x)$ becomes singular on the Fermi surface defined by $k_0 = 0$ and $e(\mathbf{k}) = 0$.

This Fermi surface depends on the kinetic energy $\epsilon(\mathbf{k})$ of the model.

For rotation invariant models, $\epsilon(\mathbf{k}) = \mathbf{k}^2/2m$ where m is some effective or "dressed" electron mass. In this case the energy is invariant under spatial rotations and the Fermi surface is simply a circle in two dimensions and a sphere in three dimensions, with radius $\sqrt{2m\mu}$. This jellium isotropic propagator is realistic in the limit of weak electron densities. We call this propagator the jellium propagator. This is the most natural model in the continuum, and in this case it is also natural to take η the ultraviolet cutoff η to be rotation invariant.[9]

Another model considered extensively is the half-filled $2d$ Hubbard model, nicknamed H_2. In this model the position variable $\mathbf{x}$ lives on the lattice $\mathbb{Z}^2$, and $\epsilon(\mathbf{k}) = \cos k_1 + \cos k_2$ so that at $\mu = 0$ the Fermi surface is a square of side size $\sqrt{2}\pi$, joining the points $(\pi, 0), (0, \pi)$ in the first Brillouin zone.

This propagator is called the Hubbard propagator.

1.2.3 Interaction, Locality

The physical interaction between conduction electrons in a solid could be very complicated; the naive Coulomb interaction is in fact subject to heavy screening, the main effective interaction being due to lattice phonons exchange and other effects. But since we are interested in long-range physics we should use a quasi-local action with rapid decay. It is a bit counterintuitive but in fact perfectly reasonable for a mathematical idealization to use a completely local interaction. This should capture all essential mathematical difficulties of the corresponding renormalization group.

There is a unique exactly local such interaction, namely

$$S_V = \lambda \int_V d^{d+1} \Big(\sum_{a \in \{\uparrow,\downarrow\}} \bar{\psi}_a(x) \psi_a(x) \Big)^2, \tag{1.104}$$

where $V := [-\beta, \beta[\times V'$ and V' is an auxiliary volume cutoff in two dimensional space, that will be sent to infinity in the thermodynamic limit. Indeed any local

---

[9] The question of whether and how to remove that ultraviolet cutoff has been discussed extensively in the literature, but we consider it as unphysical for a non-relativistic model of condensed matter, which is certainly an effective theory at best.

polynomial of higher degree is zero since Fermionic fields anti-commute. Remark it is of the same form than (1.91), with spin playing the role of color.

Hence from the mathematical point of view, in contrast with the propagator, the interesting condensed matter interaction is essentially unique.

The models with jellium propagator and such an interaction (1.104) are respectively nicknamed J_2 and J_3 in dimensions 2 and 3. The model with Hubbard propagator and interaction (1.104) is nicknamed H_2.

It is possible to interpolate continuously between H_2 and J_2 by varying the filling factor of the Hubbard model. Lattice models with next-nearest neighbor hopping are also interesting, as they are really the ones used to model the high T_c superconducting phase in cuprates, but we shall not consider them here for simplicity.

The basic new feature which changes dramatically the power counting of the theory is that the singularity of the jellium propagator is of codimension 2 in the $d+1$ dimensional space-time. Instead of changing with dimension, like in ordinary field theory, perturbative power counting is now independent of the dimension, and is the one of a just renormalizable theory. Indeed in a graph with four external legs, there are n vertices, $2n-2$ internal lines and $L = n-1$ independent loops. Each independent loop momentum gives rise to two transverse variables, for instance k_0 and $|\mathbf{k}|$ in the jellium case, and to $d-1$ inessential bounded angular variables. Hence the $2L = 2(n-1)$ dimensions of integration for the loop momenta exactly balance the $2n-2$ singularities of the internal propagators, as is the case in a just renormalizable theory.

In one spatial dimension, hence two space-time dimensions, the Fermi surface reduces to two points, and there is also no proper BCS theory since there is no continuous symmetry breaking in two dimensions (by the "Mermin–Wagner theorem"). Nevertheless the many Fermion system in one spatial dimension gives rise to an interesting non-trivial behavior, called the Luttinger liquid [59].

The marvel is that although the renormalization group now pinches a non trivial extended singularity, the locality principle still works. The two point function renormalization may change the Fermi radius, or even the shape of the Fermi surface in the non-rotation invariant case, but the corresponding flows can usually be shown to be bounded and small. But the four-point function renormalization essentially in the simplest cases changes the value of the coupling constant without changing the *local* character of the interaction (1.104). The reason for this is surprising. Indeed quasi-local graphs have internal lines which simply carry excitations farther from the Fermi surface than their external ones. Momenta close to the Fermi surface, when moved to a barycenter of the graph, should react through a non trivial phase factor. But the miracle is that the only divergent part of the main one-loop contribution, the "bubble graph" comes when the combination of the two external legs at each end carries approximately zero total momentum. This is because only then can the inner momentum integration range over the full Fermi sphere. This special configuration, with proper spin and arrows to ensure no oscillations occur, *is the Cooper pair.*

Two such external legs with approximately zero total momentum can be moved together like a single low momentum leg in an ordinary Wilsonian renormalization. This is in essence why locality for the four point function renormalization still works in condensed matter. Of course if contributions beyond one loop are taken into account, the story becomes more complicated and the renormalization group flow can in fact involve infinitely many coupling constants [60].

A still more surprising case where locality works in a new form, called Moyality, is the case of non commutative field theory on Moyal space [31]. There again the divergent subgraphs are exactly the only ones that can be renormalized through counterterms of the initial form of the Lagrangian [61]. This lead us to hope that another still generalized form of locality might hold in quantum gravity.

1.2.4 A Brief Review of Rigorous Results

What did the rigorous mathematical study of interacting Fermi systems accomplish so far? In dimension 1 there is neither superconductivity nor extended Fermi surface, and Fermion systems have been proved to exhibit Luttinger liquid behavior [59]. The initial goal of the studies in two or three dimensions was to understand the low temperature phase of these systems, and in particular to build a rigorous constructive BCS theory of superconductivity. The mechanism for the formation of Cooper pairs and the main technical tool to use (namely the corresponding 1/N expansion, where N is the number of sectors which proliferate near the Fermi surface at low temperatures) have been identified [43]. But the goal of building a completely rigorous BCS theory ab initio remains elusive because of the technicalities involved with the constructive control of continuous symmetry breaking. So the initial goal was replaced by a more modest one, still important in view of the controversies over the nature of two dimensional Fermi liquids [62], namely the rigorous control of what occurs before Cooper pair formation.

As is well known, sufficiently high magnetic field or temperature are the two different ways to break the Cooper pairs and prevent superconductivity. Accordingly two approaches were devised for the rigorous construction of interacting Fermi liquids. One is based on the use of non-parity invariant Fermi surfaces to prevent pair formation. These surfaces occur physically when generic magnetic fields are applied to two dimensional Fermi systems. In the large series of papers [58], the construction of two dimensional Fermi liquids for a wide class of non-parity invariant Fermi surfaces has been completed in great detail by Feldman, Knörrer and Trubowitz. These papers establish Fermi liquid behavior in the traditional sense of physics textbooks, namely as a jump of the density of states at the Fermi surface at zero temperature, but they do not apply to the simplest Fermi surfaces, such as circles or squares, which are parity invariant.

The other approach is based on Salmhofer's criterion, in which temperature is the cutoff which prevents pair formation. However in this case the traditional textbooks definition is inapplicable. Indeed the discontinuity never really occurs for parity-invariant models because BCS or similar phase transition are generic at low temperatures. This is called the Kohn–Luttinger effect.

Salmhofer's mathematical criterion states that an interacting system of electrons has Fermi liquid behavior if and only if

- The Schwinger functions are analytic in the coupling constant λ in a domain $|\lambda| \leq K/|\log T|$. remark that the key point is that such a domain remains *above* the low temperature phases, if K is chosen small enough.
- The self-energy as function of the momentum remains bounded uniformly together with its first and second derivatives in that domain.

The study of whether a given each model satisfies Salmhofer's criterion or not can be divided conveniently into two main steps of roughly equal difficulty, the control of convergent contributions and the renormalization of the two point functions. In dimension two the corresponding analysis has been completed for J_2, a Fermi liquid in the sense of Salmhofer, and for H_2 which is not, and is a Luttinger liquid with logarithmic corrections, according to [48–52].

Similar results similar have been also obtained for more general convex curves not necessarily rotation invariant such as those of the Hubbard model at low filling, where the Fermi surface becomes more and more circular, including an improved treatment of the four point functions leading to better constants [53, 54]. Therefore as the filling factor of the Hubbard model is moved from half-filling to low filling, we conclude that there must be a crossover from Luttinger liquid behavior to Fermi liquid behavior. This sheds light on the controversy [62] over the Luttinger or Fermi nature of two-dimensional many-Fermion systems above their critical temperature.

1.2.5 *Multiscale Analysis, Angular Sectors*

For any two-dimensional model built until now in the constructive sense, the strategy is the same. It is based on some kind of multiscale expansion, which keeps a large fraction of the theory in unexpanded determinants. The global bound on these determinant (using determinant inequalities such as Gram inequality) is much better than if the determinant was expanded into Feynman graphs which would then be bounded one by one, and the bounds summed. The bound obtained in this way would simply diverge at large order (i.e. not prove any analyticity at all in the coupling constant) simply because there are too many Feynman graphs at large order. But the divergence of a bound does not mean the divergence of the true quantity if the bound is bad. Constructive analysis, which keeps loops unexpanded is the correct way to obtain better bounds, which do prove that the true series in fact does not diverge, i.e. has a finite convergence radius in the coupling constant. This radius however shrink when the temperature goes to 0, and a good constructive analysis should establish the correct shrinking rate, which is logarithmic. This is where multiscale rather than single scale constructive analysis becomes necessary.

The basic idea of the multiscale analysis is to slice the propagator according to the size of its denominator so that the slice with index j corresponds to $|ik_0 + e(\mathbf{k})| \simeq M^{-j}$, where M is some fixed constant.

This multiscale analysis is supplemented within each scale by an angular sector analysis which further divides the Fermi slice in the directions tangential to the Fermi surface. The number of sectors should be kept as small as possible, so each sector should be as large as possible in the directions tangent to the Fermi surface in three dimensions, or to the Fermi curve in two dimensions. What limits however the size of these sectors is the curvature of the surface, so that stationary phase method could still relate the spatial decay of a propagator within a sector to its dual size in momentum space. In the case of a circle, the number of sectors at distance M^{-j} of the singularity grows therefore at least like $M^{j/2}$, hence like a power of T. However for the half-filled Hubbard model, since the curvature is "concentrated at the corners" the number of sectors grows only logarithmically at low temperature T (see Sect. 1.2.12). In one dimension there are really only two sectors since the Fermi singularity is made of two points. A logarithm is closer to a constant than to a power; this observation is the main reason for which the half-filled Hubbard model is closer to the one-dimensional Luttinger liquid than to the three dimensional Fermi liquid.

Momentum conservation rules for sectors which meet at a given vertex in general are needed to fix the correct power counting of the subgraphs of the model. In the Hubbard case at half filling, these rules are needed only to fix the correct logarithmic power counting, since the growth of sectors near the singularity is only logarithmic. In both cases the net effect in two dimensions of these conservation rules is to roughly identify two pairs of conserved sectors at any vertex, so that in each slice the model resembles an N-component vector model, where N is the number of sectors in the slice, see Sects. 1.2.7 and 1.2.14.

The multiscale renormalization group analysis of the model then consists essentially in selecting, for any graph, a tree which is a subtree in each of the quasi-local connected components of the graph according to the momentum slicing. These connected components are those for which all internal lines are farther from the Fermi surface than all external lines. The selection of this tree can be performed in a constructive manner, keeping the remaining loop fields in a determinant. The combinatoric difficulty related to the fact that a graph contains many trees is tackled by the forest formula.

1.2.6 Renormalization

Once the scale analysis has been performed, a partial expansion of the loop determinant can detect all the dangerous two and four point functions which require renormalization. A key point is that this expansion can be done without destroying the Gram bound, and the corresponding sum is not too big (this means its cardinal remains bounded by K^n (where K is a constant)) because in typical graphs there are not many two and four point subgraphs.

A particularly delicate point has to do with one and two particle irreducible expansions. Indeed Salmhofer's criterion is stated for the self-energy, i.e. the sum of all one-particle irreducible graphs for the two point function. Its study requires the correct renormalization of these contributions. Since angular sectors in a graph may vary from one propagator to the next in a graph, and since different sectors have different decays in different directions, we are in a delicate situation. In order to prove that renormalization indeed does the good that it is supposed to do, one cannot simply rely on the connectedness of these self-energy graphs, but one must use their particle irreducibility explicitly. So the proof requires a constructive particle irreducible analysis of the self-energy which is usually quite tedious. In the case of the jellium model J_2, this analysis can be performed at the level of one-particle irreducible graphs [49]. The half-filled Hubbard model, however, is more difficult.

We would like now to enter into more detail, without drowning the reader into technicalities. Hence we shall limit ourselves here to the non-perturbative analysis of connected functions in a single RG scale, which is the core mathematical problem. We compare the various models J_2, J_3 and H_2 to the toy model of Sect. 1.1.16, and explain first why sector analysis plus momentum conservation is suited to analyze the two-dimensional models but fails in three dimensions.

1.2.7 2D Jellium Model: Why Sectors Work

We claim that the J_2 model in a slice is roughly similar to the Toy Model, with dimension $d = 3$, provided the momentum slice is divided along angular directions into regions called *sectors*, which play the role of colors.

The naive estimate on the slice propagator is (using integration by parts)

$$|C_j(x,y)| \leq M^{-j} e^{-[M^{-j}|x-y|]^{1/2}} \tag{1.105}$$

(using Gevrey cutoffs f_j to get fractional exponential decay). The prefactor M^{-j} corresponds to the volume of integration of the slice, M^{-2j}, divided by the slice estimate of the denominator, M^{-j}. This is much worse than the factor $M^{-3j/2}$ that would be needed.

But the situation improves if we cut the Fermi slice into *smaller pieces* (called sectors). Suppose we divide the jth slice into M^j sectors, each of size roughly M^{-j} in all three directions.

A sector propagator $C^{j,a}$ has now prefactor M^{-2j} corresponding to the volume of integration of the sector M^{-3j} divided by the slice estimate of the denominator, M^{-j}. Using integration by parts and Gevrey cutoffs f_{ja} for fractional power decay we get without too much effort the bound

$$|C_{j,ab}(x,y)| \leq \delta_{ab} M^{-2j} e^{-[M^{-j}|x-y|]^{1/2}}. \tag{1.106}$$

But since $N = M^j$

$$M^{-2j} = \frac{M^{-3j/2}}{\sqrt{N}} \tag{1.107}$$

so that this bound is identical to that of the toy model.

It remains just to explain why the interaction of the model is approximatively of the vector type. This is because of the momentum conservation rule at every vertex.

In two dimensions a rhombus (i.e. a closed quadrilateral whose four sides have equal lengths) is a parallelogram. Hence an approximate rhombus should be an approximate parallelogram.

Momentum conservation $\delta(p_1 + p_2 + p_3 + p_4)$ at each vertex follows from translation invariance of J_2. Hence p_1, p_2, p_3, p_4 form a quadrilateral. For j large we have $|p_k| \simeq \sqrt{2M\mu}$, hence the quadrilateral is an approximate rhombus. Hence the four sectors to which p_1, p_2, p_3 and p_4 should be roughly equal two by two (parallelogram condition).

It means that the interaction is roughly of the color (or Gross–Neveu) type with respect to these angular sectors:

$$\Big(\sum_a \bar{\psi}_a \psi_a\Big)\Big(\sum_b \bar{\psi}_b \psi_b\Big). \tag{1.108}$$

In fact this "rhombus rule" is not fully correct for almost degenerate rhombuses. The correct statement is

Lemma 1.4. *Fix $m \in \mathbf{Z}^3$. The number of 4-tuples $\{S_1, \cdots S_4\}$ of sectors for which there exist $k_i \in \mathbb{R}^2$, $i = 1, \cdots, 4$ satisfying*

$$k_i' \in S_i, \quad |k_i - k_i'| \leq \text{const}\, M^{-j}, \quad i = 1, \cdots, 4 \tag{1.109}$$

and

$$|k_1 + \cdots + k_4| \leq \text{const}\,(1 + |m|)\, M^{-j} \tag{1.110}$$

is bounded by

$$\text{const}(1 + |m|)^2 M^{2j} \{1 + j\}. \tag{1.111}$$

The $1 + j$ factor is special to dimension 2 and is the source of painful technical complications which were developed by Feldman, Magnen, Trubowitz and myself.

The solution uses in fact $M^{j/2}$ *anisotropic* angular sectors, which are longer in the tangential direction (of length $M^{-j/2}$). The corresponding propagators still have dual spatial decay because the sectors are still approximately flat.

Ultimately the conclusion is unchanged: the radius of convergence of J_2 in a slice is independent of the slice index j.

The following theorem summarizes the results of [48, 49]:

Theorem 1.2. *The radius of convergence of the jellium two-dimensional model perturbative series for any thermodynamic function is at least $c/|\log T|$, where T is the temperature and c some numerical constant. As T and λ jointly tend to 0 in this* *domain, the self-energy and its first two momentum derivatives remain uniformly bounded so that the model is a Fermi liquid in the sense of Salmhofer.*

1.2.8 3D Jellium Model: Why Sectors Fail

In the three dimensional jellium model, sectors and Gram's bound fail by a full power per vertex!

There is indeed no rhombus rule in $d = 3$. A closed quadrilateral with equal sides is not a parallelogram because it can be non-planar; hence it is obtained by rotating half of a planar parallelogram around the diagonal by an arbitrary *twisting angle*. Therefore the jellium model interaction is *not* of the vector type. More precisely the analog of Lemma 1.4 is, with similar notations

Lemma 1.5. *The number of* 4*-tuples* $\{S_1, \cdots S_4\}$ *of sectors for which there exist* $k_i \in \mathbb{R}^3$, $i = 1, \cdots, 4$ *satisfying*

$$k_i' \in S_i, \quad |k_i - k_i'| \leq \text{const}\, M^{-j}, \quad i = 1, \cdots, 4 \tag{1.112}$$

and

$$|k_1 + \cdots + k_4| \leq \text{const}\,(1 + |m|)\, M^{-j} \tag{1.113}$$

is bounded by

$$\text{const}(1 + |m|)^3 M^{5j}. \tag{1.114}$$

Remark the absence of the log factor that was present in $d = 2$. But for $d = 3$ we find M^{5j} 4-tuples which correspond to the choice of two sectors ($M^{2j} \times M^{2j}$) and of one angular twist M^j. The power counting corresponds to M^{-3j} per sector propagator. Two propagators pay for one vertex integration (M^{4j}) and one sector choice (M^{2j}) but there is *nothing to pay for the angular twist.* Going to anisotropic sectors is possible but there remains still in this case a $M^j/2$ twist factor. After many years of effort, we concluded that the sector method and Gram bound apparently cannot be improved to do better. Hence although J_3 is expected to be a Fermi liquid in the sense of Salmhofer, new methods have to be developed to treat it constructively [63–65].

1.2.9 The Hadamard Method in *x*-Space

The idea of using Hadamard's inequality in x-space to overcome the constructive power counting problem of Fermions in three dimensions took Jacques Magnen and myself 4 years of continuous hard work, with dozens and dozens of various failed trials, from 1991 to 1995 [63]. Another 4 or 5 years took place to fine-tune this idea for the multidimensional case with Disertori [64]. Another 10 years have passed to find a momentum-conserving version, which should at last allow for the proof of Salmhofer's criterion in the three dimensional jellium model [65].

Let us describe the main idea in an informal way.

We remember that the naive estimate (1.105) is far from sufficient for correct power counting. But we also know from perturbative power counting in momentum space that the theory should be just renormalizable. Therefore the j-slice propagator C^j should behave more as the one of an infrared ϕ_4^4 theory, hence should be bounded by

$$KM^{-2j}e^{-[M^{-j}|x-y|]^{1/2}}. \tag{1.115}$$

In fact this is not totally correct, because it can be shown that the jth slice propagator at almost coinciding points, hence at $|x-y| \simeq 0$ is not bounded by M^{-2j}. Still (1.115) is *correct at typical distances* for the J_3 propagator in the jth slice. We know from eg (1.105) that these typical space-time distances should be $|x-y| \simeq M^j$.

Indeed integrating over angles on the Fermi sphere leads to an *additional* $1/|x-y|$ decay, because

$$\int_0^{\pi} \sin\theta d\theta d\phi e^{i\cos\theta|x-y|} = \sin|x-y|/|x-y|. \tag{1.116}$$

Hence for typical distances $|x-y| \simeq M^j$ the propagator indeed obeys the improved estimate

$$|C_j(x,y)_{|x-y|\simeq M^j}| \leq KM^{-2j}e^{-[M^{-j}|x-y|]^{1/2}}. \tag{1.117}$$

The problem is that this bound is wrong at small distances. However at small distances there is a bonus, namely the corresponding integration volumes are smaller. Another related problem is that if we use the Gram bound (1.85) to bound a determinant with all the matrix elements $C_j(x_p - y_q)$ corresponding to large distances $|x_p - y_q|$, we still loose the improvement (1.117), because the L^2 norms in (1.85) will correspond again to propagators at coinciding points!

The solution is to use the Hadamard bounds (1.86)–(1.87) because they conserve the decay of the typical propagators.

Remember however, as seen conveniently in (1.88) that Hadamard bounds consume the $1/n!$ symmetry factor for n vertices. Hence we can no longer use the explicit tree formula and the method of [41, 42].

But we can still use the standard old-fashioned cluster expansion between cubes. Summarizing:

- Just renormalizable power counting is recovered for the main part of the theory if we use Hadamard's bound rather than Gram's bound.
- The factor $n!$ is lost in the Hadamard bound at order n; this forces us to rely on the on-canonical tool of cluster expansion between cubes.

Still, this solves the constructive problem only for the main part of the propagator, the one at typical distances $|x-y| \simeq M^j$. However at smaller distances there is a bonus, namely the volume factors for spatial integration are also smaller.

It turns out that the problem of *smaller than typical* distances can be solved with an auxiliary superrenormalizable decomposition

$$C^j = \sum_{k=0}^{j} C^{jk} \tag{1.118}$$

of the propagator. Roughly speaking $C^{j,k}$ corresponds to $|x-y| \simeq M^k$, It means that even the single slice theory in $d \geq 3$ is a non-trivial theory that contains a rather non-trivial renormalization, like the one of ϕ^4_3. This renormalization can be analyzed by means of the auxiliary scales. The solution is in fact more complicated that what we sketch here, and has to take into account the anisotropy between the space and the imaginary time variables [63, 64].

The use of non-canonical lattices of cubes in this method is the signal that we have probably still not found the optimal constructive treatment of J_3. Recently we found a better decomposition that should allow to check Salmhofer's criterion for $J3$ [65]. However in its present stage it will still use a non-canonical cluster expansion between cubes. It would be interesting to find a solution such as the loop vertex expansion that solves the constructive problem in a truly canonical way.

1.2.10 2D Hubbard Model

The Hubbard model lives on the square lattice $\mathbb{Z}^2$, so that the three dimensional vector $x = (x_0, \mathbf{x})$ is such that $\mathbf{x} = (n_1, n_2) \in \mathbb{Z}^2$. From now on we write v_1 and v_2 for the two components of a vector $\mathbf{v}$ along the two axis of the lattice.

At half-filling and finite temperature T, the Fourier transform of the propagator of the Hubbard model is:

$$\hat{C}_{ab}(k) = \delta_{ab} \frac{1}{ik_0 - e(\mathbf{k})}, \qquad e(\mathbf{k}) = \cos k_1 + \cos k_2 , \tag{1.119}$$

where $a, b \in \{\uparrow, \downarrow\}$ are the spin indices. The vector $\mathbf{k}$ lives on the two-dimensional torus $\mathbb{R}^2/(2\pi\mathbb{Z})^2$. Hence the real space propagator is

$$C_{ab}(x) = \frac{1}{(2\pi)^2 \beta} \sum_{k_0} \int_{-\pi}^{\pi} dk_1 \int_{-\pi}^{\pi} dk_2 \, e^{ikx} \, \hat{C}_{ab}(k) . \tag{1.120}$$

Recall that $|k_0| \geq \pi/\beta \neq 0$ hence the denominator in $C(k)$ again can never be 0 at non zero temperature. This is why the temperature provides a natural infrared cut-off. When $T \to 0$ (which means $\beta \to \infty$) k_0 becomes a continuous variable, the discrete sum becomes an integral, and the corresponding propagator $C_0(x)$ becomes singular on the Fermi surface defined by $k_0 = 0$ and $e(\mathbf{k}) = 0$. This Fermi surface is a square of side size $\sqrt{2}\pi$ (in the first Brillouin zone) joining the corners $(\pm\pi, 0), (0, \pm\pi)$. We call this square the Fermi square, its corners and faces are called the Fermi faces and corners. Considering the periodic boundary conditions, there are really four Fermi faces, but only two Fermi corners.

In the following to simplify notations we will write:

$$\int d^3k \equiv \frac{1}{\beta}\sum_{k_0}\int d^2k, \quad \int d^3x \equiv \frac{1}{2}\int_{-\beta}^{\beta} dx_0 \sum_{\mathbf{x}\in\mathbb{Z}^2} . \tag{1.121}$$

In determining the spatial decay we recall that by anti-periodicity

$$C(x) = f(x_0, \mathbf{x}) := \sum_{m\in\mathbb{Z}} (-1)^m \, C_0\left(x_0 + \frac{m}{T}, \mathbf{x}\right). \tag{1.122}$$

where C_0 is the propagator at $T = 0$. Indeed the function f is anti-periodic and its Fourier transform is the right one.

The interaction of the Hubbard model is again (1.104):

$$S_V = \lambda \int_V d^3x \, (\sum_a \bar{\psi}\psi)^2(x) \,, \tag{1.123}$$

where $V := [-\beta, \beta] \times V'$ and V' is an auxiliary finite volume cutoff in two dimensional space that will be sent later to infinity.

1.2.11 Scale Analysis

The theory has a natural lattice spatial cutoff. To implement the renormalization group analysis, we introduce as usually a compact support function $u(r) \in \mathscr{C}_0^\infty(\mathrm{R})$ (it is convenient to choose it to be Gevrey of order $\alpha < 1$ so as to ensure fractional exponential decrease in the dual space) which satisfies:

$$u(r) = 0 \quad \text{for } |r| > 2, \; u(r) = 1 \quad \text{for } |r| < 1. \tag{1.124}$$

With this function, given a constant $M \geq 2$, we can construct a partition of unity

$$1 = \sum_{i=0}^{\infty} u_i(r) \;\; \forall r \neq 0,$$
$$u_0(r) = 1 - u(r), \; u_i(r) \; = \; u(M^{2(i-1)}r) - u(M^{2i}r) \text{ for } i \geq 1. \tag{1.125}$$

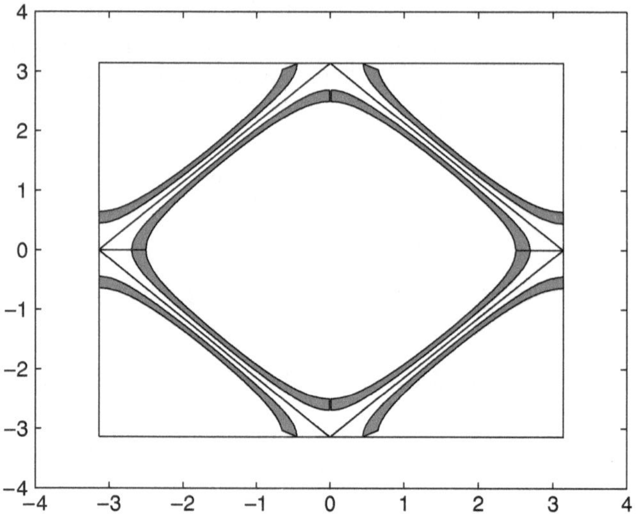

Fig. 1.6 A single slice of the renormalization group

The propagator is then divided into slices according to this partition

$$C(k) = \sum_{i=0}^{\infty} C_i(k) \tag{1.126}$$

where

$$C_i(k) = C(k) u_i[k_0^2 + e^2(\mathbf{k})]. \tag{1.127}$$

(indeed $k_0^2 + e^2(\mathbf{k}) \geq T^2 > 0$).

In a slice of index i the cutoffs ensure that the size of $k_0^2 + e^2(\mathbf{k})$ is roughly M^{-2i}. More precisely in the slice i we must have

$$M^{-2i} \leq k_0^2 + e^2(\mathbf{k}) \leq 2M^2 M^{-2i}. \tag{1.128}$$

The corresponding domain is a three dimensional volume whose section through the $k_0 = 0$ plane is the shaded region pictured in Fig. 1.6.

Remark that at finite temperature, the propagator C_i vanishes for $i \geq i_{max}(T)$ where $M^{i_{max}(T)} \simeq 1/T$ (more precisely $i_{max}(T) = E(\log \frac{M\sqrt{2}}{\pi T} / \log M)$, where E is the integer part), so there is only a finite number of steps in the renormalization group analysis.

Let us state first a simple result, for a theory whose propagator is only C_i, hence corresponding to a generic step of the renormalization group:

Theorem 1.3. *The Schwinger functions of the theory with propagator C_i and interaction (1.123) are analytic in λ in a disk of radius R_i which is at least c/i*

for a suitable constant c:

$$R_i \geq c/i. \tag{1.129}$$

The rest of this section is devoted to the definitions and properties of H_2 sectors, their scaled decay and momentum conservation rules. As discussed already this result is a first step towards the rigorous proof [51,52] that H_2 is not a Fermi liquid in the sense of Salmhofer.

1.2.12 Sectors

This section is adapted from [50].

The "angular" analysis is completely different from the jellium case. We remark first that in a slice, $k_0^2 + e^2(\mathbf{k})$ is of order M^{-2i}, but this does not fix the size of $e^2(\mathbf{k})$ itself, which can be of order M^{-2j} for some $j \geq i$. In order for sectors defined in momentum space to correspond to propagators with dual decay in direct space, it is essential that their length in the tangential direction is not too big, otherwise the curvature is too strong for stationary phase methods to apply. This was discussed first in [43]. This leads us to study the curve $(\cos k_1 + \cos k_2)^2 = M^{-2j}$ for arbitrary $j \geq i$. We can by symmetry restrict ourselves to the region $0 \leq k_1 \leq \pi/2, k_2 > 0$. It is then easy to compute the curvature radius of that curve, which is

$$R = \frac{(\sin^2 k_1 + \sin^2 k_2)^{3/2}}{|\cos k_1 \sin^2 k_2 + \cos k_2 \sin^2 k_1|}. \tag{1.130}$$

We can also compute the distance $d(k_1)$ to the critical curve $\cos k_1 + \cos k_2 = 0$, and the width $w(k_1)$ of the band $M^{-j} \leq |\cos k_1 + \cos k_2| \leq \sqrt{2}M.M^{-j}$. We can then easily check that

$$d(k_1) \simeq w(k_1) \simeq \frac{M^{-j}}{M^{-j/2} + k_1}, \tag{1.131}$$

$$R(k_1) \simeq \frac{k_1^3 + M^{-3j/2}}{M^{-j}}, \tag{1.132}$$

where $f \simeq g$ means that on the range $0 \leq k_1 \leq \pi/2$ we have inequalities $cf \leq g \leq df$ for some constants c and d.

Defining the anisotropic length

$$l(k_1) = \sqrt{w(k_1)R(k_1)} \simeq M^{-j/2} + k_1, \tag{1.133}$$

the condition in [43] is that the sector length should not be bigger than that anisotropic length. This leads to the idea that k_1 or an equivalent quantity should be sliced according to a geometric progression from 1 to $M^{-j/2}$ to form the angular sectors in this model.

For symmetry reasons it is convenient to introduce a new orthogonal but not normal basis in momentum space (e_+, e_-), defined by $e_+ = (1/2)(\pi, \pi)$ and $e_- = (1/2)(-\pi, \pi)$. Indeed if we call (k_+, k_-) the coordinates of a momentum k in this basis, the Fermi surface is given by the simple equations $k_+ = \pm 1$ or $k_- = \pm 1$. This immediately follows from the identity

$$\cos k_1 + \cos k_2 = 2\cos(\pi k_+/2)\cos(\pi k_-/2). \tag{1.134}$$

(Note however that the periodic b.c. are more complicated in that new basis). Instead of slicing $e(\mathbf{k})$ and k_1, it is then more symmetric to slice directly $\cos(\pi k_+/2)$ and $\cos(\pi k_-/2)$.

Guided by these considerations we introduce the partition of unity

$$1 = \sum_{s=0}^{i} v_s(r); \begin{cases} v_0(r) = 1 - u(M^2 r) & \\ v_s = u_{s+1} & \text{for} 1 \leq s \leq i-1 \\ v_i(r) = u(M^{2i} r) & \end{cases} \tag{1.135}$$

and define

$$C_i(k) = \sum_{\sigma=(s_+,s_-)} C_{i,\sigma}(k) \tag{1.136}$$

where

$$C_{i,\sigma}(k) = C_i(k) v_{s_+}[\cos^2(\pi k_+/2)]\, v_{s_-}[\cos^2 \pi k_-/2)]. \tag{1.137}$$

We remark that using (1.128) in order for $C_{i,\sigma}$ not to be 0, we need to have $s_+ + s_- \geq i-2$. We define the "depth" $l(\sigma)$ of a sector to be $l = s_+ + s_- - i + 2$.

To get a better intuitive picture of the sectors, we remark that they can be classified into different categories:

- The sectors (0,i) and (i,0) are called the middle-face sectors
- The sectors (s,i) and (i,s) with $0 < s < i$ are called the face sectors
- The sector (i,i) is called the corner sector
- The sectors (s,s) with $(i-2)/2 \leq s < i$ are called the diagonal sectors
- The others are the general sectors

Finally the general or diagonal sectors of depth 0 for which $s_+ + s_- = i - 2$ are called border sectors.

If we consider the projection onto the (k_+, k_-) plane, taking into account the periodic b.c. of the Brillouin zone, the general and diagonal sectors have eight connected components, the face sectors have four connected components, the middle face sectors and the corner sector have two connected components. In the three dimensional space-time, if we neglect the discretization of the Matsubara frequencies, these numbers would double except for the border sectors.

It can then be proved that these sectors obey indeed direct-space decay dual to their momentum size support [50].

1.2.13 Support Properties

If $C_{i,\sigma}(k) \neq 0$, the momentum k must obey the following bounds:

$$|k_0| \leq \sqrt{2}MM^{-i} \tag{1.138}$$

$$\begin{cases} M^{-1} \leq |\cos(\pi k_\pm/2)| \leq 1 & \text{for } s_\pm = 0\,, \\ M^{-s_\pm - 1} \leq |\cos(\pi k_\pm/2)| \leq \sqrt{2}M^{-s_\pm} & \text{for } 1 \leq s_\pm \leq i-1\,, \\ |\cos(\pi k_\pm/2)| \leq \sqrt{2}M^{-i} & \text{for } s_\pm = i\,. \end{cases} \tag{1.139}$$

In the support of our slice in the first Brillouin zone we have $|k_+| < 2$ and $|k_-| < 2$ (this is not essential but the inequalities are strict because $i \geq 1$). It is convenient to associate to any such component $k_\pm$ a kind of "fractional part" called $q_\pm$ defined by $q_\pm = k_\pm - 1$ if $k_\pm \geq 0$ and $q_\pm = k_\pm + 1$ if $k_\pm < 0$, so that $0 \leq |q_\pm| \leq 1$. Then the bounds translate into

$$\begin{cases} 2/\pi M \leq |q_\pm| \leq 1 & \text{for } s_\pm = 0\,, \\ 2M^{-s_\pm}/\pi M \leq |q_\pm| \leq \sqrt{2}M^{-s_\pm} & \text{for } 1 \leq s_\pm \leq i-1\,, \\ |q_\pm| \leq \sqrt{2}M^{-i} & \text{for } s_\pm = i\,. \end{cases} \tag{1.140}$$

1.2.14 Momentum Conservation Rules at a Vertex

Let us consider that the four momenta k_1, k_2, k_3, k_4, arriving at a given vertex v belong to the support of the four sectors σ_1, σ_2, σ_3, σ_4, in slices i_1, i_2, i_3, i_4. In Fourier space the vertex (1.123) implies constraints on the momenta. Each spatial component of the sum of the four momenta must be an integer multiple of 2π in the initial basis, and the sum of the four Matsubara frequencies must also be zero.

In our tilted basis (e_+, e_-), this translates into the conditions:

$$k_{1,0} + k_{2,0} + k_{3,0} + k_{4,0} = 0\,, \tag{1.141}$$

$$k_{1,+} + k_{2,+} + k_{3,+} + k_{4,+} = 2n_+\,, \tag{1.142}$$

$$k_{1,-} + k_{2,-} + k_{3,-} + k_{4,-} = 2n_-\,, \tag{1.143}$$

where n_+ and n_- must have identical parity.

We want to rewrite the two last equations in terms of the fractional parts q_1, q_2, q_3 and q_4.

Since an even sum of integers which are ± 1 is even, we find that (1.142) and (1.143) imply

$$q_{1,+} + q_{2,+} + q_{3,+} + q_{4,+} = 2m_+ \, , \tag{1.144}$$

$$q_{1,-} + q_{2,-} + q_{3,-} + q_{4,-} = 2m_- \, , \tag{1.145}$$

with m_+ and m_- integers. Let us prove now that except in very special cases, these integers must be 0. Since $|q_{j,\pm}| \leq 1$, $|m_\pm| \leq 2$. But $|q_{j,\pm}| = 1$ is possible only for $s_{j,\pm} = 0$. Therefore $|m_\pm| = 2$ implies $s_{j,\pm} = 0 \; \forall j$. Now suppose e.g. $|m_+| = 1$. Then $s_{j,+}$ is 0 for at least two values of j. Indeed for $s_{j,\pm} \neq 0$ we have $|q_{j,\pm}| \leq \sqrt{2}M^{-1}$, and assuming $3\sqrt{2}M^{-1} < 1$, (1.144) could not hold.

We have therefore proved

Lemma 1.6. *$m_+ = 0$ unless $s_{j,+}$ is 0 for at least two values of j, and $m_- = 0$ unless $s_{j,-}$ is 0 for at least two values of j.*

Let us analyze in more detail (1.144) and (1.145) for $|m_+| = |m_-| = 0$. Consider e.g. (1.144). By a relabeling we can assume without loss of generality that $s_{1,+} \leq s_{2,+} \leq s_{3,+} \leq s_{4,+}$ Then either $s_{1,+} = i_1$ or $s_{1,+} < i_1$, in which case combining (1.144) and (1.140) we must have:

$$3\sqrt{2}M^{-s_{2,+}} \geq 2M^{-s_{1,+}}/\pi M \, , \tag{1.146}$$

which means

$$s_{2,+} \leq s_{1,+} + 1 + \frac{\log(3\pi/\sqrt{2})}{\log M} \, . \tag{1.147}$$

This implies

$$|s_{2,+} - s_{1,+}| \leq 1 \tag{1.148}$$

if $M > 3\pi/\sqrt{2}$, which we assume from now on.

The conclusion is:

Lemma 1.7. *If $m_\pm = 0$, either the smallest index $s_{1,\pm}$ coincides with its scale i_1, or the two smallest indices among $s_{j,\pm}$ differ by at most one unit.*

Now we can summarize the content of both Lemmas in a slightly weaker but simpler lemma:

Lemma 1.8.

(A) (Single slice case)
The two smallest indices among $s_{j,+}$ for $j = 1,2,3,4$ differ by at most one unit, and the two smallest indices among $s_{j,-}$ for $j = 1,2,3,4$ differ by at most one unit.

(B) Multislice case
The two smallest indices among $s_{j,+}$ for $j = 1,2,3,4$ differ by at most one unit or the smallest one, say $s_{1,+}$ must coincide with its scale i_1, which must

then be strictly smaller than the three other scales i_2, i_3 and i_4. Exactly the same statement holds independently for the minus direction.

This lemma allows to check that again the single scale analysis works and leads to a radius of convergence independent of the slice [50].

1.2.15 Multiscale Analysis

The half-filling point is very convenient since the particle-hole exact symmetry at this point ensures that there is no flow for the Fermi surface itself.

Using the sector decomposition and the momentum conservation we find that power counting of the 2D Hubbard model is essentially similar to one dimensional case with logarithmic corrections [51].

Although there is no real divergence of the self-energy (the associated counterterm is zero thanks to the particle hole symmetry of the model at half-filling) one really needs a two-particle and one-vertex irreducible constructive analysis to establish the necessary constructive bounds on the self-energy and its derivatives [51]. For parity reasons, the self-energy graphs of the model are in fact not only one-particle irreducible but also two particle and one vertex irreducible, so that this analysis is possible.

This analysis leads to the explicit construction of three line-disjoint paths for every self-energy contribution, in a way compatible with constructive bounds. On top of that analysis, another one which is scale-dependent is performed: after reduction of some maximal subsets provided by the scale analysis, two vertex-disjoint paths are selected in every self-energy contribution. This construction allows to improve the power counting for two point subgraphs, exploiting the particle-hole symmetry of the theory at half-filling, and leads to the desired analyticity result.

Finally an upper bound for the self energy second derivative is combined with a lower bound for the explicit leading self energy Feynman graph [52]. This completes the proof that the Hubbard model violates Salmhofer's criterion, hence is not a Fermi liquid, in contrast with the jellium two dimensional model. More precisely the following theorem summarizes the results of [50–52].

Theorem 1.4. *The radius of convergence of the Hubbard model perturbative series at half-filling is at least $c/|\log T|$, where T is the temperature and c some numerical constant. As T and λ jointly tend to 0 in this domain, the self-energy of the model does not display the properties of a Fermi liquid in the sense of Salmhofer, since the second derivative is not uniformly bounded.*

Acknowledgements I thank J. Magnen, M. Disertori, M. Smerlak and L. Gouba for contributing various aspects of this work.

References

1. J. Glimm, A. Jaffe, *Quantum Physics. A Functional Integral Point of View* (McGraw Hill, New York, 1981)
2. R. Feynman, A. Hibbs, *Quantum Mechanics and Path Integrals* (McGraw Hill, New York, 1965)
3. M. Peskin, D.V. Schroeder (Contributor), *An Introduction to Quantum Field Theory* (Perseus Publishing, New York, 1995)
4. C. Itzykson, J.-B. Zuber, *Quantum Field Theory* (McGraw Hill, New York, 1980)
5. P. Ramond, *Field Theory* (Addison-Wesley, Boston, 1994)
6. J. Zinn-Justin, *Quantum Field Theory and Critical Phenomena* (Oxford University Press, Oxford, 2002)
7. C. Itzykson, J.M Drouffe, *Statistical Field Theory*, vols. 1 and 2 (Cambridge University Press, Cambridge, 1991)
8. G. Parisi, *Statistical Field Theory* (Perseus Publishing, New York, 1998)
9. V. Rivasseau, *From Perturbative to Constructive Renormalization* (Princeton University Press, Princeton, 1991)
10. J.P. Eckmann, J. Magnen, R. Sénéor, Decay properties and Borel summability for the Schwinger functions in $P(\phi)_2$ theories. Comm. Math. Phys. **39**, 251 (1975)
11. J. Magnen, R. Sénéor, Phase space cell expansion and Borel summability for the Euclidean ϕ_3^4 theory. Comm. Math. Phys. **56**, 237 (1977)
12. A. Sokal, An improvement of Watson's theorem on Borel summability. J. Math. Phys. **21**, 261 (1980)
13. F. Bergeron, G. Labelle, P. Leroux, *Combinatorial Species and Tree-like Structures* (Cambridge University Press, Cambridge, 1998)
14. V. Rivasseau et al., (Eds.) *Constructive Physics*. Lecture Notes in Physics, vol. 446 (Springer, Berlin, 1995)
15. V. Rivasseau, Constructive field theory and applications: Perspectives and open problems. J. Math. Phys. **41**, 3764 (2000)
16. V. Rivasseau, Constructive matrix theory. J. High Energ. Phys. **09**, 008 (2007) [arXiv:hep-ph/0706.1224]
17. J. Magnen, V. Rivasseau, Constructive field theory without tears. Ann. Henri Poincaré **9**, 403–424 (2008), math/ph/0706.2457
18. R. Gurau, J. Magnen, V. Rivasseau, Tree quantum field theory. Ann. Henri Poincaré **10**(5), 867–891 (2009) [arXiv:0807.4122]
19. V. Rivasseau, Constructive field theory in zero dimension. Adv. Math. Phys. **2009** (2009). Article ID 180159 [arXiv:0906.3524]
20. V. Rivasseau, Z. Wang, How are Feynman graphs resummed by the loop vertex expansion [arXiv:1006.4617]
21. T. Krajewski, V. Rivasseau, A. Tanasa, Z. Wang, Topological graph polynomials and quantum field theory. Part I: Heat kernel theories. J. Noncommut. Geom. **4**, 29–82 (2010) [arXiv:0811.0186]
22. K. Wilson, Renormalization group and critical phenomena, II Phase space cell analysis of critical behavior. Phys. Rev. B **4**, 3184 (1974)
23. K. Hepp, *Théorie de la renormalisation* (Springer, Berlin, 1969)
24. N. Bogoliubov, O. Parasiuk, Acta Math. **97**, 227 (1957)
25. W. Zimmermann, Convergence of Bogoliubov's method for renormalization in momentum space. Comm. Math. Phys. **15**, 208 (1969)
26. J. Polchinski, Nucl. Phys. B **231**, 269 (1984)
27. C. Kopper, in *Renormalization Theory Based on Flow Equations*. Progress in Mathematics, vol. 251 (Springer, Berlin, 2007), pp. 161–174
28. M. Bergère, Y.M.P. Lam, Bogoliubov-Parasiuk theorem in the α-parametric representation. J. Math. Phys. **17**, 1546 (1976)

29. M. Bergère, J.B. Zuber, Renormalization of Feynman amplitudes and parametric integral representation. Comm. Math. Phys. **35**, 113 (1974)
30. C. de Calan, V. Rivasseau, Local existence of the Borel transform in Euclidean ϕ_4^4. Comm. Math. Phys. **82**, 69 (1981)
31. H. Grosse, R. Wulkenhaar, Renormalization of ϕ^4-theory on noncommutative $\mathbb{R}^4$ in the matrix base. Comm. Math. Phys. **256**(2), 305–374 (2005) [hep-th/0401128]
32. H. Grosse, R. Wulkenhaar, The beta-function in duality-covariant noncommutative ϕ^4-theory. Eur. Phys. J. **C35**, 277–282 (2004) [hep-th/0402093]
33. M. Disertori, V. Rivasseau, Two and three loops beta function of non commutative Phi44 theory. Eur. Phys. J. C **50**, 661 (2007) [arXiv:hep-th/0610224]
34. M. Disertori, R. Gurau, J. Magnen, V. Rivasseau, Vanishing of beta function of non commutative Φ_4^4 theory to all orders. Phys. Lett. B **649**, 95–102 (2007) [arXiv:hep-th/0612251]
35. J. Feldman, in *Renormalization Group and Fermionic Functional Integrals*. CRM Monograph Series, vol. 16 (1999) (AMS, Providence)
36. N. Nakanishi, *Graph Theory and Feynman integrals* (Gordon and Breach, New York, 1971)
37. K. Symanzik, in *Local Quantum Theory*, ed. by R. Jost (Varenna, 1968) (Academic, New York, 1969), p. 285
38. A. Abdesselam, The Grassmann-Berezin calculus and theorems of the matrix-tree type. Adv. Appl. Math. **33**(1), 51–70 (2004)
39. D. Brydges, T. Kennedy, Mayer expansions and the Hamilton-Jacobi equation. J. Stat. Phys. **48**, 19 (1987)
40. A. Abdesselam, V. Rivasseau, in *Trees, Forests and Jungles: A Botanical Garden for Cluster Expansions*. Constructive Physics, Lecture Notes in Physics, vol. 446 (Springer, Berlin, 1995) [arXiv:hep-th/9409094]
41. A. Lesniewski, Effective action for the Yukawa_2 quantum field theory. Comm. Math. Phys. **108**, 437 (1987)
42. A. Abdesselam, V. Rivasseau, Explicit fermionic tree expansions. Lett. Math. Phys. **44**(1), 77–88 (1998) [arXiv:cond-mat/9712055]
43. J. Feldman, J. Magnen, V. Rivasseau, E. Trubowitz, An infinite volume expansion for many Fermion Green's functions. Helv. Phys. Acta **65**, 679 (1992)
44. V. Rivasseau, Z. Wang, Loop vertex expansion for ϕ^{2k} theory in zero dimension. J. Math. Phys. **51** (2010) [arXiv:1003.1037/092304]
45. G. Benfatto, G. Gallavotti, Perturbation theory of the Fermi surface in a quantum liquid. A general quasi-particle formalism and one dimensional systems. J. Stat. Phys. **59**, 541 (1990)
46. J. Feldman, E. Trubowitz, Perturbation theory for many fermions systems. Helv. Phys. Acta **63**, 156 (1990)
47. J. Feldman, E. Trubowitz, The flow of an electron-phonon system to the superconducting state. Helv. Phys. Acta **64**, 213 (1991)
48. M. Disertori, V. Rivasseau, Interacting Fermi liquid in two dimensions at finite temperature. Part I: Convergent attributions. Comm. Math. Phys. **215**, 251 (2000)
49. M. Disertori, V. Rivasseau, A rigorous proof of fermi liquid behavior for jellium two-dimensional interacting Fermions. Phys. Rev. Lett. **85**, 361 (2000)
50. V. Rivasseau, The two dimensional Hubbard model at half-filling: I. Convergent contributions. J. Stat. Phys. **106**, 693–722 (2002)
51. S. Afchain, J. Magnen, V. Rivasseau, Renormalization of the 2-point function of the Hubbard model at half-filling. Ann. Henri Poincaré **6**, 399 (2005)
52. S. Afchain, J. Magnen, V. Rivasseau, The Hubbard model at half-filling. Part III: The lower bound on the self-energy. Ann. Henri Poincaré **6**, 449 (2005)
53. G. Benfatto, A. Giuliani, V. Mastropietro, Low temperature analysis of two dimensional Fermi systems with symmetric Fermi surface. Ann. Henri Poincaré **4**, 137 (2003)
54. G. Benfatto, A. Giuliani, V. Mastropietro, Fermi liquid behavior in the 2D Hubbard model at low temperatures. Ann. Henri Poincaré **7**, 809 (2006)
55. A. Giuliani, V. Mastropietro, Rigorous construction of ground state correlations in graphene: Renormalization of the velocities and Ward identities. Phys. Rev. B **79**, 201403(R) (2009)

56. A. Giuliani, V. Mastropietro, M. Porta, Lattice gauge theory model for graphene. Phys. Rev. B **82**, 121418(R) (2010)
57. M. Salmhofer, Comm. Math. Phys. **194**, 249 (1998)
58. J. Feldman, H. Knörrer, E. Trubowitz, A two dimensional Fermi Liquid. Comm. Math. Phys. **247**, 1–319 (2004); Rev. Math. Phys. **15**(9), 949–1169 (2003). Papers accessible at http://www.math.ubc.ca/feldman/fl.html
59. G. Benfatto, G. Gallavotti, *Renormalization Group* (Princeton University Press, Princeton, 1995)
60. J. Feldman, J. Magnen, V. Rivasseau, E. Trubowitz, Two dimensional many Fermion systems as vector models. Europhys. Lett. **24**, 521 (1993)
61. R. Gurau, J. Magnen, V. Rivasseau, F. Vignes-Tourneret, Renormalization of non-commutative Φ_4^4 field theory in x space. Comm. Math. Phys. **267**, 515 (2006)
62. P.W. Anderson, Luttinger liquid behavior of the normal metallic state of the 2D Hubbard model. Phys. Rev. Lett. **64**, 1839–1841 (1990)
63. J. Magnen, V. Rivasseau, A single scale infinite volume expansion for three-dimensional many Fermion Green's functions. Math. Phys. Electron. J. **1**(3) (1995)
64. M. Disertori, J. Magnen, V. Rivasseau, Interacting Fermi liquid in three dimensions at finite temperature. Part I: Convergent contributions. Ann. Henri Poincaré **2**, 733–806 (2001)
65. M. Disertori, J. Magnen, V. Rivasseau, Parametric cutoffs for interacting Fermi liquid, [arXiv:1105.4138] http://arxiv.org/abs/1105.4138

Chapter 2
Cold Quantum Gases and Bose–Einstein Condensation

Robert Seiringer

2.1 Introduction

Bose–Einstein condensation (BEC) in cold atomic gases was first achieved experimentally in 1995 [1, 6]. After initial failed attempts with spin-polarized atomic hydrogen, the first successful demonstrations of this phenomenon used gases of rubidium and sodium atoms, respectively. Since then there has been a surge of activity in this field, with ingenious experiments putting forth more and more astonishing results about the behavior of matter at very cold temperatures. BEC has now been achieved by more than a dozen different research groups working with gases of different types of atoms. Literally thousands of scientific articles, concerning both theory and experiment, have been published in recent years.

The theoretical investigation of BEC goes back much further, and even predates the modern formulation of quantum mechanics. It was investigated in two papers by Einstein [9] in 1924 and 1925, respectively, following up on a work by Bose [3] on the derivation of Planck's radiation law. Einstein's result, in its modern formulation, can be found in any textbook on quantum statistical mechanics, and was concerned with ideal, i.e., non-interacting gases.

The understanding of BEC in the presence of interparticle interactions poses a formidable challenge to mathematical physics. Some progress was made in the last 10 years or so, and the purpose of these lecture notes is to explain part of what was achieved and how it is related to the actual experiments on cold gases. The content of these notes is naturally strongly biased towards the work of the author and no attempt of completeness is made.

R. Seiringer (✉)
Department of Mathematics and Statistics, McGill University, 805 Sherbrooke Street West, Montreal, QC, Canada H3A 2K6
e-mail: rseiring@math.mcgill.ca

V. Rivasseau et al., *Quantum Many Body Systems*, Lecture Notes in Mathematics 2051, DOI 10.1007/978-3-642-29511-9_2,

2.2 Quantum Many-Body Systems

2.2.1 *The Hamiltonian*

For a system of N particles, the Hamiltonian typically takes the form

$$H_N = \sum_{i=1}^{N} \frac{p_i^2}{2m_i} + V(x_1, \ldots, x_N)$$

where m_i is the mass of particle i, $p_i \in \mathbb{R}^3$ and $x_i \in \mathbb{R}^3$ are its momentum and position, respectively, and V denotes the total potential energy. Classically, H is a function on phase space, but in quantum mechanics it becomes a linear operator with the substitution $p_j = -i\hbar\nabla_j$. We shall choose units such that $\hbar = 1$ from now on.

The Hamiltonian H_N is a linear operator on Hilbert space, which is a suitable subspace of $L^2(\mathbb{R}^{3N})$, the square integrable functions of N variables $x_i \in \mathbb{R}^3$. Only a subspace is relevant physically, since for two identical bosons, say i and j, there is the symmetry requirement

$$\psi(x_1, \ldots, x_i, \ldots, x_j, \ldots, x_N) = \psi(x_1, \ldots, x_j, \ldots, x_i, \ldots x_N)$$

For fermions, there is an additional minus sign, i.e., the wave function is antisymmetric with respect to exchange of coordinates. For simplicity of notation, we ignore here internal degrees of freedom of the particles, like spin, but these could easily be taken into account by adding to the coordinates x_i these additional parameters.

The form of the potential energy depends on the physical system under consideration. Typically, it is a sum of various terms, containing one-particle potentials of the form $\sum_i W(x_i)$, corresponding to an external force, two-body interaction potentials $\sum_{i<j} W(x_i, x_j)$ for pairwise interaction, or even some more complicated interactions involving more than two particles at the same time. We will usually assume that

$$V(x_i, \ldots, x_N) = \sum_{i=1}^{N} W(x_i) + \sum_{1\leq i<j\leq N} v(|x_i - x_j|).$$

2.2.2 *Quantities of Interest*

Given the Hamiltonian H_N of a quantum systems, there are many questions one can try to address. The first one might be concerning its ground state energy, i.e., the lowest values of the spectrum, which we denote by

$$E_0(N) = \inf \operatorname{spec} H_N$$

If $E_0(N)$ is an eigenvalue, the corresponding ground state wave function ψ_0 is determined by Schrödinger's equation $H_N\psi_0 = E_0(N)\psi_0$.

More generally, if the system is at some positive temperature $T > 0$, one would like to compute the free energy of the system, given by

$$F = -T \ln \operatorname{Tr} e^{-H_N/T}$$

We choose units such that Boltzmann's constant equals 1, and shall often write $T = 1/\beta$. The trace is over the physical Hilbert space, of course, respecting symmetry constraints arising form the indistinguishability of particles. The equilibrium state at temperature T is the Gibbs state

$$\rho_\beta = e^{-\beta(H-F)}$$

It is normalized to have $\operatorname{Tr}\rho_\beta = 1$. For large particle number, it is usually hopeless to try to calculate ρ_β directly, but one will try to investigate properties of the reduced n-particle density matrices, obtained by taking the partial trace of ρ_β over $N - n$ variables.

It is often convenient not to fix the particle number N, but rather work in the grand-canonical ensemble, where one takes a certain average over the number of particles in the system. For simplicity, consider a system of just one species of particles. The N-particle Hilbert space, $\mathscr{H}_N$, is then the set of square-integrable functions that are either totally symmetric or antisymmetric under permutations, depending on whether the particles are bosons or fermions.

In the grand-canonical ensemble, one has as Hilbert space the Fock space

$$\mathscr{F} = \bigoplus_{N=0}^{\infty} \mathscr{H}_N$$

Here, $\mathscr{H}_0 = \mathbb{C}$ by definition, and the corresponding vector is called the vacuum vector. As Hamiltonian on Fock space one simply takes

$$H = \bigoplus_{N=0}^{\infty} H_N$$

with H_N the N-particle Hamiltonian. Typically, $H_0 = 0$, i.e., the vacuum has zero energy.

For $\mu \in \mathbb{R}$, the grand canonical potential is defined as

$$J = -T \ln \operatorname{Tr}_{\mathscr{F}} e^{-\beta(H-\mu N)}$$

where N denotes the number operator, i.e,

$$N = \bigoplus_{n=0}^{\infty} n$$

Since H is particle number conserving, we can also write this as

$$J = -T \ln \sum_{N \geq 0} z^N \operatorname{Tr}_{\mathcal{H}_N} e^{-\beta H_N}$$

where $z = e^{\beta\mu}$ is called the fugacity.

The grand-canonical Gibbs state is

$$\rho_{\beta,\mu} = e^{-\beta(H-\mu N-J)}$$

The chemical potential μ is adjusted to achieve a given average particle number $\langle N \rangle$. The latter equals

$$\langle N \rangle = \operatorname{Tr} N \rho_{\beta,\mu} = -\frac{\partial}{\partial\mu} J$$

2.2.3 Creation and Annihilation Operators on Fock Space

On Fock space $\mathcal{F}$, a particularly useful concept are the creation and annihilation operators $a^\dagger(f)$ and $a(f)$, with $f \in \mathcal{H}_1$, the one-particle Hilbert space. For any $N \geq 0$, we have

$$a^\dagger(f) : \mathcal{H}_N \to \mathcal{H}_{N+1}$$

i.e., it creates a particle. Likewise, $a(f)$ annihilates a particle, i.e.,

$$a(f) : \mathcal{H}_N \to \mathcal{H}_{N-1}$$

Explicitly, they are defined as follows. If ψ_N is an N-particle wave function in $\mathcal{H}_N$,

$$\big(a^\dagger(f)\psi_N\big)(x_1,\ldots,x_{N+1}) = \frac{1}{\sqrt{N+1}} \sum_{i=1}^{N+1} f(x_i)\psi_N(x_1,\ldots,\not{x}_i,\ldots,x_{N+1})$$

and

$$\big(a(f)\psi_N\big)(x_1,\ldots,x_{N-1}) = \sqrt{N} \int_{\mathbb{R}^3} \bar{f}(x_N)\psi_N(x_1,\ldots,x_{N-1},x_N)dx_N$$

This definition works for bosons, for fermions one has to introduce the appropriate (-1) factors to preserve the antisymmetry of the wave functions. One readily checks that these operators satisfy $a(f)^\dagger = a^\dagger(f)$, i.e., $a^\dagger(f)$ is the adjoint of $a(f)$, as well as the canonical (anti-)commutation relations

$$[a(f), a^\dagger(g)] = \langle f|g\rangle\ ,\ [a(f), a(g)] = 0\ ,\ [a^\dagger(f), a^\dagger(g)] = 0$$

Here, $[\cdot, \cdot]$ denotes the usual commutator $[A, B] = AB - BA$ for bosons, while it is the anticommutator $[A, B] = AB + BA$ for fermions.

Consider now a typical many-body Hamiltonian containing one- and two-body terms. For h a one-body operator and W a two-body operator, the N-particle Hamiltonian is thus of the form

$$H_N = \sum_{i=1}^{N} h_i + \sum_{1 \leq i < j \leq N} W_{ij}$$

where the subscripts indicate what particles the operator acts on. Using creation and annihilation operators, the Fock space Hamiltonian $H = \bigoplus_{N \geq 0} H_N$ can conveniently be written as

$$H = \sum_{i,j} \langle \varphi_i | h | \varphi_j \rangle \, a_i^\dagger a_j + \frac{1}{2} \sum_{i,j,k,l} \langle \varphi_i \otimes \varphi_j | W | \varphi_k \otimes \varphi_l \rangle \, a_i^\dagger a_j^\dagger a_k a_l$$

where $\{\varphi_i\}$ is an orthonormal basis of $\mathcal{H}_1$, and $a_i^\dagger = a^\dagger(\varphi_i)$, $a_i = a(\varphi_i)$. A possible choice of the basis $\{\varphi_i\}$ is to diagonalize h, i.e.,

$$\langle \varphi_i | h | \varphi_j \rangle = e_i \delta_{ij}$$

The number operator N is simply $N = \sum_i a_i^\dagger a_i$.

In terms of the creation and annihilation operators, the reduced n-particle density matrices $\gamma^{(n)}$ of a state on Fock space are defined via the expectation values

$$\langle f_1 \otimes \cdots \otimes f_n | \gamma^{(n)} | g_1 \otimes \cdots \otimes g_n \rangle = \langle a^\dagger(g_1) \ldots a^\dagger(g_n) a(f_n) \ldots a(f_1) \rangle$$

Since product functions span the whole n-particle space, this defines $\gamma^{(n)}$ uniquely. For a state with a fixed particle number, the definition agrees with the previous definition in the canonical ensemble using partial traces (except for an overall normalization factor).

2.2.4 *Ideal Quantum Gases*

Consider now an ideal quantum system without interactions. The N-particle Hamiltonian is simply $H_N = \sum_{i=1}^{N} h_i$, where, for example,

$$h = \frac{1}{2m} p^2 = -\frac{1}{2m} \Delta$$

on the cube $[0, L]^3$ with appropriate boundary conditions. In particular, we assume that h has discrete spectrum. Let us denote the eigenvalues of h by e_i,

$$e_0 \leq e_1 \leq e_2 \leq \dots$$

On Fock space, we then have

$$H = \sum_{i \geq 0} e_i a_i^\dagger a_i$$

and also

$$\beta H - \mu N = \sum_{i \geq 0} \epsilon_i a_i^\dagger a_i$$

with $\epsilon_i = \beta e_i - \mu$.

We wish to calculate

$$\ln \operatorname{Tr} e^{-\sum_i \epsilon_i a_i^\dagger a_i}$$

The spectrum of $\sum_i \epsilon_i a_i^\dagger a_i$ is of the form $\sum_i \epsilon_i n_i$, with $n_i \in \{0, 1, 2, \dots\}$ for bosons, and $n_i \in \{0, 1\}$ for fermions. Summing over all possible occupation numbers is the same as summing over all eigenstates, hence we have

$$\operatorname{Tr} e^{-\sum_i \epsilon_i a_i^\dagger a_i} = \prod_i \sum_n e^{-\epsilon_i n} = \prod_i \begin{cases} (1 - e^{-\epsilon_i})^{-1} & \text{bosons} \\ 1 + e^{-\epsilon_i} & \text{fermions} \end{cases}$$

For bosons, we have to assume that $\epsilon_i > 0$ for all i for the geometric series to converge. In particular

$$\ln \operatorname{Tr} e^{-\sum_i \epsilon_i a_i^\dagger a_i} = \sum_i \mp \ln(1 \mp e^{-\epsilon_i})$$

where $-$ is for bosons and $+$ for fermions.

Consider now an ideal gas in a cubic box of side length L, with periodic boundary conditions. The spectrum of $p^2 = -\Delta$ equals

$$\left(\frac{2\pi}{L}\right)^2 \left(n_x^2 + n_y^2 + n_z^2\right)$$

with $(n_x, n_y, n_z) \in \mathbb{Z}^3$. The corresponding eigenstates are the plane waves $e^{ip \cdot x}$, with $p \in (\frac{2\pi}{L}\mathbb{Z})^3$. The grand canonical potential (which equals the negative of the pressure times the volume in this case) thus equals

$$J = \pm T \sum_{p \in (\frac{2\pi}{L}\mathbb{Z})^3} \ln\left(1 \mp e^{-\beta(p^2 - \mu)}\right)$$

where we set the mass of the particles equal to $1/2$ for simplicity.

For bosons, we have to assume that $\mu < 0$. This is not really a restriction, however, as any particle number can be achieved even for negative μ. In fact, the

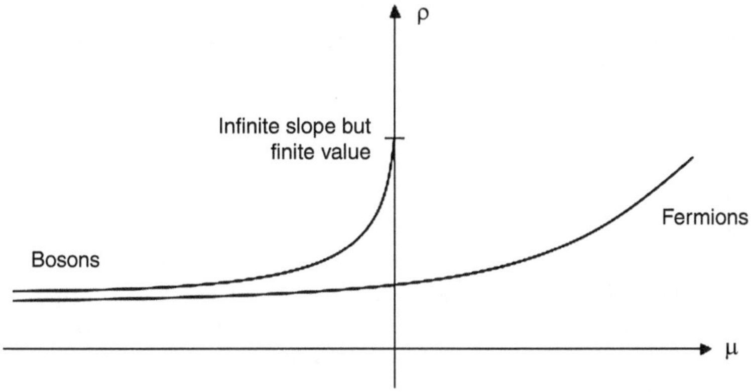

Fig. 2.1 Particle density of an ideal quantum gas at infinite volume, as a function of the chemical potential μ

average particle number equals

$$\langle N \rangle = -\frac{\partial}{\partial \mu} J = \sum_{p \in (\frac{2\pi}{L}\mathbb{Z})^3} \underbrace{\frac{1}{e^{\beta(p^2-\mu)} \mp 1}}_{\langle a_p^\dagger a_p \rangle}$$

Here, the summands are just $\langle a_p^\dagger a_p \rangle$, the average occupation number of momentum p. As μ varies within $(-\infty, 0)$ (for bosons), and $(-\infty, \infty)$ (for fermions), clearly $\langle N \rangle$ varies between 0 and $+\infty$.

We now perform a thermodynamic limit $L \to \infty$. The sum over p can then be interpreted as a Riemann sum for the corresponding integral. In fact,

$$\frac{1}{L^3} \sum_{p \in (\frac{2\pi}{L}\mathbb{Z})^3} \longrightarrow \frac{1}{(2\pi)^3} \int_{\mathbb{R}^3} dp$$

as $L \to \infty$. The thermodynamic pressure of the system is thus

$$P = -\lim_{L \to \infty} \frac{J}{L^3} = \mp \frac{T}{(2\pi)^3} \int_{\mathbb{R}^3} \ln\left(1 \mp e^{-\beta(p^2-\mu)}\right) dp$$

and the average density equals (Fig. 2.1).

$$\rho = \lim_{L \to \infty} \frac{\langle N \rangle}{L^3} = \frac{1}{(2\pi)^3} \int_{\mathbb{R}^3} \frac{1}{e^{\beta(p^2-\mu)} \mp 1} dp. \tag{2.1}$$

Let us know restrict our attention to the bosonic case, where there is a minus sign in the denominator in (2.1). Notice that the density stays bounded as $\mu \to 0$! I.e.,

$$\rho_c(\beta) := \lim_{\mu \nearrow 0} \rho = \frac{1}{(2\pi)^3} \int_{\mathbb{R}^3} \frac{1}{e^{\beta p^2} - 1} dp < \infty \tag{2.2}$$

since the integrand behaves like $|p|^{-2}$ for small p, which is integrable in three dimensions.

What is happening here? Recall that μ has to be chosen as to fix the density ρ and, hence, has to depend on L, in general. If $\rho < \rho_c(\beta)$, then $\mu(L) \to \mu < 0$ in the thermodynamic limit. But when $\rho \geq \rho_c(\beta)$, $\mu(L)$ has to tend to zero as $L \to \infty$. In this case, the limits $L \to \infty$ and $\mu \to 0$ must be taken simultaneously and, in particular, do not commute.

In fact, if $\rho > \rho_c(\beta)$, then μ is asymptotically equal to

$$\mu = \left(-\beta L^3(\rho - \rho_c(\beta))\right)^{-1} \quad \text{as } L \to \infty$$

For this value of μ, we see that

$$\lim_{L\to\infty} \frac{1}{L^3} \langle a_0^\dagger a_0 \rangle = \lim_{L\to\infty} \frac{1}{L^3} \frac{1}{e^{-\beta\mu} - 1} = \rho - \rho_c(\beta)$$

That is, the zero momentum state is occupied by a macroscopic fraction of all the particles. This phenomenon is called *Bose–Einstein Condensation* (BEC). It occurs for $\rho > \rho_c(\beta)$, i.e., for ρ bigger than the critical density or, equivalently, for

$$T < T_c(\rho) = \frac{4\pi}{\zeta(3/2)^{2/3}} \rho^{2/3}$$

since $\rho_c(\beta) = \zeta(3/2)(4\pi)^{-3/2}\beta^{-3/2}$. Here, ζ denotes the Riemann zeta function

$$\zeta(z) = \sum_{k\geq 1} k^{-z}$$

I.e., BEC occurs below a critical temperature $T_c(\rho)$.

We note that only the zero momentum mode is macroscopically occupied, and the other occupations are much smaller. The smallest positive eigenvalue of the Laplacian equals $(2\pi/L)^2$, and

$$\frac{1}{e^{\beta(2\pi/L)^2} - 1} \sim L^2 \ll L^3 \quad \text{for large } L.$$

BEC represents a phase transition in the usual sense, namely that the thermodynamic functions exhibit a non-analytic behavior. Consider, for instance, the free energy, which is given in a standard way as the Legendre transform of the pressure.

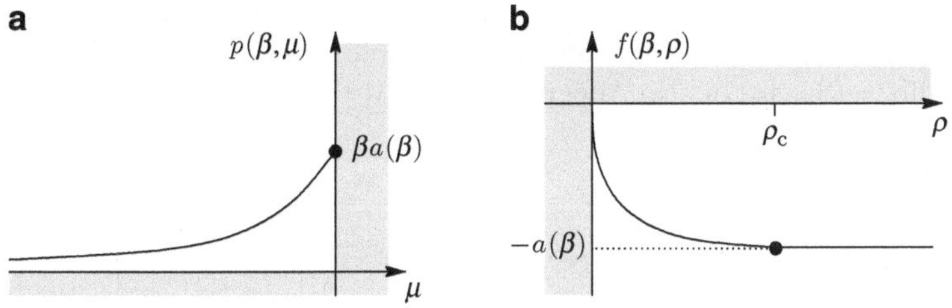

Fig. 2.2 The pressure and the free energy of the ideal Bose gas in three dimensions

Specifically, the free energy per unit volume equals

$$f(\beta,\rho) = \mu\rho + \frac{T}{(2\pi)^3}\int_{\mathbb{R}^3} \ln\left(1 - e^{-\beta(p^2-\mu)}\right) dp \tag{2.3}$$

where μ is determined by (2.1) if $\rho < \rho_c(\beta)$, and $\mu = 0$ if $\rho \geq \rho_c(\beta)$. In the latter case, we see that $f(\beta,\rho)$ does not actually depend on ρ, and is constant for $\rho > \rho_c(\beta)$. In particular, f is not analytic. Intuitively, what is happening as one increases the density beyond $\rho_c(\beta)$ is that all additional particles occupy the zero momentum mode and hence do not contribute to the energy or the entropy, hence also not to the free energy (Fig. 2.2).

We conclude this section by noting that the grand-canonical ensemble is somewhat unphysical for the ideal Bose gas for $\rho > \rho_c(\beta)$, because of large particle number fluctuations. One readily computes that

$$\langle n_p(n_p - 1)\rangle = 2\langle n_p\rangle^2 \tag{2.4}$$

for any p, where $n_p = a_p^\dagger a_p$ denotes the number of particles with momentum p. Note that (2.4) is independent of β and μ. It can be easily checked from the explicit form of the Gibbs state, or follows from Wick's rule, for instance.

The factor 2 on the right side of (2.4) is crucial. It means that the variance of the occupation number is of the same order is its value. In particular, if there is a macroscopic occupation, also the variance is macroscopic! By summing over p, we also see that

$$\langle N(N-1)\rangle = \langle N\rangle^2 + \sum_p \langle n_p\rangle^2$$

The last term of is of the order $\langle N\rangle^2$ if and only if $\rho > \rho_c(\beta)$. That is, there are macroscopically large particle number fluctuations in this case. For real, interacting systems, such fluctuation will typically be suppressed and this problem is not expected to occur.

The macroscopic particle number fluctuations in particular mean that there is not a full equivalence of ensembles for the ideal Bose gas. Nevertheless, working in the canonical ensemble will produce the same free energy (2.3) and the same occupation numbers $\langle n_p \rangle$ in the thermodynamic limit. In particular, the conclusion concerning BEC remains the same in the canonical ensemble, although the analysis is much more tedious. (See [4] or also Appendix B of [33].)

2.3 BEC for Interacting Systems

2.3.1 *The Criterion for BEC*

Recall the definition of the one-particle density matrix γ. For $\langle \cdot \rangle$ a state on Fock space,

$$\langle g|\gamma|f\rangle = \langle a^\dagger(f)a(g)\rangle$$

Note that γ is a positive trace class operator on the one-particle Hilbert space $\mathscr{H}_1$, with

$$\operatorname{Tr}\gamma = \sum_i \langle\varphi_i|\gamma|\varphi_i\rangle = \left\langle \sum_i a_i^\dagger a_i \right\rangle = \langle N\rangle$$

This definition applies to any state on Fock space, not only thermal equilibrium states. In particular, one can also consider states of definite particle number, and hence recover the definition for the canonical ensemble.

According to a criterion by Penrose and Onsager [26], BEC is said to occur if γ has an eigenvalue of the order of $\langle N\rangle$ of large $\langle N\rangle$. The corresponding eigenfunction is called the *condensate wave function.*

Since this definition involves large particle numbers $\langle N\rangle$, it refers, strictly speaking, not to a single state but rather a sequence of states for larger and larger system size. When one speaks about the occurrence of BEC one hence always has to specify how various parameters depend on this size.

The standard case where one would like to understand BEC is a translation invariant system at given inverse temperature β and chemical potential μ, in the limit that the system size L tends to infinity. In this case, BEC means that

$$\lim_{L\to\infty} \frac{1}{L^3} \underbrace{\sup_{\|f\|=1} \langle a^\dagger(f)a(f)\rangle_{\beta,\mu}}_{\text{largest eigenvalue of } \gamma} > 0.$$

For translation invariant systems, γ is also translation invariant and hence has an integral kernel of the form

$$\gamma(x-y) = \frac{1}{L^3} \sum_{p\in\left(\frac{2\pi}{L}\mathbb{Z}\right)^3} \gamma_p\, e^{ip(x-y)}$$

where γ_p denote the eigenvalues of γ. Moreover, for a Hamiltonian of the form

$$H = \sum_{i=1}^{N} p_i^2 + V(x_1, \dots, x_N)$$

it is also true that γ is positivity improving, meaning that it has a strictly positive integral kernel $\gamma(x-y)$, and hence $\gamma_0 > \gamma_p$ for all $p \neq 0$. Hence the largest eigenvalue is always associated to the constant eigenfunction, and BEC can only occur in the zero momentum mode.

BEC is extremely hard to establish rigorously. In fact, the only known case of an interacting, translation invariant Bose gas where BEC has been proved in the standard thermodynamic limit is the hard-core lattice gas. For completeness, we shall briefly describe it in the following section.

2.3.2 The Hard-Core Lattice Gas

For a lattice gas, one replaces the continuous configuration space $\mathbb{R}^3$ of a particle by the cubic lattice $\mathbb{Z}^3$. That is, the one-particle Hilbert space $\mathscr{H}_1$ becomes $\ell^2(\mathbb{Z}^3)$ instead of $L^2(\mathbb{R}^3)$. Other types of lattices are possible, of course, but we restrict our attention to the simple cubic one for simplicity. The appropriate generalization of the Laplacian operator on $\mathbb{R}^3$ is the discrete Laplacian

$$(\Delta\psi)(x) = \sum_e (\psi(x+e) - \psi(x)),$$

where the sum is over unit vectors e pointing to the nearest neighbors on the lattice.

We assume that the interaction between particles takes place only on a single site, and that it is sufficiently strong to prevent any two particles from occupying the same site. In this sense, these are hard-core particles. I.e., the interaction energy is zero if all particles occupy different sites, and $+\infty$ otherwise.

Since there is at most one particle at a site x, we can represent the creation and annihilation operators of a particle at a site x as 2×2 matrices

$$a_x^\dagger = \begin{pmatrix} 0 & 1 \\ 0 & 0 \end{pmatrix}, \qquad a_x = \begin{pmatrix} 0 & 0 \\ 1 & 0 \end{pmatrix}$$

where the vector $\binom{1}{0}_x$ refers to the state where x is occupied, and $\binom{0}{1}_x$ to the state where x is empty. Also

$$n_x = a_x^\dagger a_x = \begin{pmatrix} 1 & 0 \\ 0 & 0 \end{pmatrix}$$

In other words, the appropriate Fock space for this system becomes

$$\mathscr{F} = \bigotimes_{x \in [0,L)^3 \cap \mathbb{Z}^3} \mathbb{C}^2_x$$

and the Hamiltonian (minus the chemical potential times N) equals

$$H = -\sum_{\langle x,y \rangle} a^\dagger_x a_y - \mu \sum a^\dagger_x a_x.$$

Here, $\langle x, y \rangle$ stands for nearest neighbor pairs on the lattice. The diagonal terms in the discrete Laplacian have been dropped for simplicity, as they can be absorbed into the chemical potential μ.

Note that Fock space is finite dimensional! Moreover, the Hamiltonian looks extremely simple, as it is quadratic in the $a^\dagger_x$ and a_x. However, these are not the usual creation and annihilation operators anymore, as they do not satisfy the canonical commutation relations. In fact,

$$[a_x, a^\dagger_x] = 1 - 2n_x$$

and

$$a_x a^\dagger_x + a^\dagger_x a_x = 1$$

Therefore, at a given site, the system looks like it is fermionic, but for different sites the operators still commute, as appropriate for bosons.

To gain some intuition about the behavior of the system, it is instructive to rewrite the Hamiltonian H in terms of spin operators. Recall that for a spin $1/2$ particle, the three components of the spin are represented by the $1/2$ times the Pauli matrices, i.e., by

$$S^1 = \frac{1}{2}\begin{pmatrix} 0 & 1 \\ 1 & 0 \end{pmatrix}, \quad S^2 = \frac{1}{2}\begin{pmatrix} 0 & -i \\ i & 0 \end{pmatrix}, \quad S^3 = \frac{1}{2}\begin{pmatrix} 1 & 0 \\ 0 & -1 \end{pmatrix}$$

If we define, as usual, the spin raising and lowering operators by $S^\pm = S^1 \pm iS^2$, we see that $a^\dagger$ corresponds to S^+, a to S^-, and n to $S^+S^- = S^3 + 1/2$. Hence, up to an irrelevant constant

$$H = -\sum_{\langle x,y \rangle} S^+_x S^-_y - \mu \sum_x S^3_x$$

This is known as the (spin $1/2$) XY model. The chemical potential plays the role of an external magnetic field (in the 3-direction).

The following theorem establishes the existence of BEC in this system at small enough temperature, for a particular value of the chemical potential.

Theorem 1 (Dyson et al. [8]) *For $\mu = 0$ and T small enough,*

$$\lim_{L\to\infty} \frac{1}{L^3} \left\langle \underbrace{\left(L^{-3/2}\sum_x S_x^+\right)}_{a^\dagger_{p=0}} \underbrace{\left(L^{-3/2}\sum_x S_x^-\right)}_{a_{p=0}} \right\rangle > 0 \tag{2.5}$$

We note that $\mu = 0$ corresponds to half-filling, i.e., $\langle N \rangle = L^3/2$, and there is a particle-hole symmetry, implying there are half as many particles as there are lattice sites, on average. The result has only been proved in this special case and it is not known how to extend it to $\mu \neq 0$.

Equation (2.5) can be rewritten as

$$\lim_{L\to\infty} \frac{1}{L^6} \sum_{x,y} \left(\langle S_x^+ S_y^- \rangle - \underbrace{\langle S_x^+ \rangle \langle S_y^- \rangle}_{=0} \right) > 0$$

since $\langle S_x^+ \rangle = \langle S_y^- \rangle = 0$ by rotation symmetry of H in the 1–2 plane. What this says is that, on average, $\langle S_x^+ S_y - \rangle - \langle S_x^+ \rangle \langle S_y^- \rangle > 0$ even though x and y are macroscopically separated. This property is known as long-range order, and is equivalent to ferromagnetism of the spin system. On average, the value of a spin is zero, but the spins tend to align in the sense that if a spin at some point x points in some directions, all other spins tend to align in the same direction.

The proof of Theorem 1 relies crucially on a special property of the system known as reflection positivity. It extends earlier results by Fröhlich, Simon and Spencer [10] on classical spin systems, where this property was first used to proof the existence of phase transitions. Reflection positivity holds only in the case of particle-hole symmetry, i.e., $\mu = 0$, and hence the proof is restricted to this particular case.

2.4 Dilute Bose Gases

2.4.1 The Model

In this section, we return to the description of Bose gases in continuous space. For simplicity, let us consider a system of just one species of particles, with pairwise interaction potential. In practice, the gas will consist of atoms, but we can treat them as point particles as long as the temperature and the density are low enough so that excitations of the atoms are rare. The atoms will behave like bosons if the number of neutrons in their nucleus is even, since then they will have an integer total spin.

The Hamiltonian for such a system is

$$H_N = \sum_{i=1}^{N} -\Delta_i + \sum_{1\leq i<j\leq N} v\left(|x_i - x_j|\right) \tag{2.6}$$

where we again choose units such that $\hbar = 1$ and $m = 1/2$. The particles are confined to a cubic box of side length L, and appropriate boundary conditions have to be chosen to make $-\Delta$ a self-adjoint operator. Usually these are Dirichlet boundary conditions (rigid walls) or periodic boundary conditions (torus).

We assume that the interaction is of short range, by which we mean that

$$\int_{|x|\geq R} v(|x|)dx < \infty$$

for some $R \geq 0$. In other words, v should be integrable at infinity. Locally it can be very strong, however. A typical example is a system of hard spheres of diameter R_0, where

$$v(|x|) = \begin{cases} +\infty & \text{if } |x| \leq R_0 \\ 0 & \text{if } |x| > R_0 \end{cases}$$

The interaction has to be sufficiently repulsive to ensure that the system is a gas for low temperatures and densities. In particular, there should be no bound states of any kind. This is certainly the case if $v(|x|) \geq 0$ for all particle separations $|x|$, which we shall assume henceforth.

So far it has not been possible to prove the existence of BEC (in the usual thermodynamic limit) for such a system, even at low density and for weak interaction v.[1] So our goals have to be more modest here. Let us first investigate the ground state energy of the system, i.e.,

$$E_0(N) = \inf \operatorname{spec} H_N$$

We will be particularly interested in large systems, i.e., in the thermodynamic limit

$$\left.\begin{array}{l} L \to \infty \\ N \to \infty \end{array}\right\} \text{ with } \rho = \frac{N}{L^3} \text{ fixed}$$

At low density, one might expect that the ground state energy is mainly determined by two-particle collisions, and hence

$$E_0(N) \approx \frac{N(N-1)}{2} E_0(2)$$

[1] It is possible to prove an upper bound on the critical temperature, however. That is, one can establish the absence of BEC for large enough temperature, see [31].

That is, the energy should approximately equal the energy of just two particles in a large box, multiplied by the total number of pairs of particles. We shall compute $E_0(2)$ in the following.

2.4.2 The Two-Particle Case

Consider now two particles in a large cubic box of side length L. Ignoring boundary conditions, the two-particle wave function will be of the form $\psi(x_1, x_2) = \phi(x_1 - x_2)$. Hence

$$\frac{\langle\psi|H_2|\psi\rangle}{\langle\psi|\psi\rangle} = \frac{\int \left(2|\nabla\phi(x)|^2 + v(|x|)|\phi(x)|^2\right) dx}{\int |\phi(x)|^2 dx}$$

since the center-of-mass integration yields L^3 both in the numerator and the denominator. Moreover, since the interaction is short range we can assume that $\phi(x)$ tends to a constant for large $|x|$, and we can take the constant to be 1 without loss of generality. Hence $\int |\phi|^2 = L^3$ to leading order in L.

Definition 1 *The* scattering length *a is defined to be*

$$a = \frac{1}{8\pi} \inf_{\phi} \left\{ \int_{\mathbb{R}^3} \left(2|\nabla\phi(x)|^2 + v(|x|)|\phi(x)|^2\right) dx \; : \; \lim_{|x|\to\infty} \phi(x) = 1 \right\} \tag{2.7}$$

Note that integrability of $v(|x|)$ at infinity is equivalent to the scattering length a being finite.

With this definition and the preceding arguments, we see that

$$E_0(2) \approx \frac{8\pi a}{L^3}$$

for large L. Hence we expect that

$$E_0(N) \approx \frac{N(N-1)}{2} E_0(2) \approx 4\pi N a \rho \quad \text{for small } \rho = N/L^3$$

We will investigate the validity of this formula in the next subsection.

We note that the Euler–Lagrange equation for the minimization problem (2.7) is

$$-2\Delta\phi(x) + v(|x|)\phi(x) = 0.$$

This is the zero-energy scattering equation. Asymptotically, as $|x| \to \infty$, the solution is of the form

$$\phi(x) \approx 1 - \frac{a}{|x|}$$

with a the scattering length of v. This is easily seen to be an equivalent definition of a, but we shall find the variational characterization (2.7) to be more useful in the following.

2.4.3 The Ground State Energy of a Dilute Gas

Consider the ground state energy per particle, $E_0(N)/N$, of the Hamiltonian (2.6) in the thermodynamic limit

$$e_0(\rho) = \lim_{N\to\infty} \frac{1}{N} E_0(N) \quad \text{with } L^3 = N/\rho$$

Based on the discussion above, we expect that

$$e_0(\rho) \approx 4\pi a\rho$$

for small density ρ. This is in fact true.

Theorem 2 (Dyson [7], Lieb and Yngvason [20])

$$e_0(\rho) = 4\pi a\rho(1 + o(1))$$

with $o(1)$ *going to zero as* $\rho \to 0$.

The upper bound was proved by Dyson in 1957 [7] using a variational calculation. He also proved a lower bound, which was 14 times too small, however. The correct lower bound was finally shown in 1998 by Lieb and Yngvason [20].

We remark that the low density limit is very different from the perturbative weak-coupling limit. In fact, at low density the energy of a particle is very small compared with the strength of v. The interaction potential is hence very strong but short range. First order perturbation theory, in fact, would predict a ground state energy of the form

$$e_0(\rho) = \frac{\rho}{2} \int v(|x|)dx$$

This is strictly bigger than $4\pi a\rho$, as can be seen from the variational principle (2.7); $(8\pi)^{-1} \int v$ is the first order Born approximation to the scattering length a.

The proof of Theorem 2 is too lengthy to be given here in full detail, but we shall explain the main ideas. For the upper bound, one can use the variational principle, which says that

$$E_0(N) \le \frac{\langle\Psi|H_N|\Psi\rangle}{\langle\Psi|\Psi\rangle} \tag{2.8}$$

for any Ψ. As a trial function that captures the right two-body physics, one could try a function of the form

$$\Psi(x_1,\ldots,x_N) = \prod_{1\leq i<j\leq N} \phi(x_i - x_j).$$

The computation of the corresponding energy turns out to be rather tricky, however. One of the reasons for this is that both numerator and denominator on the right side of (2.8) are exponentially small in the particle number N, and hence cancellations have to be taken into account very carefully. Dyson in fact used a slightly different form of the trial function, and his computation of the upper bound fills several pages.

Before explaining the main ideas in the lower bound by Lieb and Yngvason, let us give some intuition as to why this is a hard problem. It is related to the relevant length scales in the system. Since the energy per particle is of the order of $a\rho$, the associated uncertainty principle length ℓ, obtained by setting this energy equal to ℓ^{-2}, equals

$$\ell \sim \frac{1}{\sqrt{a\rho}}$$

At low density ρ, this is

$$\ell \sim \frac{1}{\sqrt{a\rho}} \quad \gg \quad \underbrace{\rho^{-1/3}}_{\text{mean interparticle separation}} \quad \gg \quad \underbrace{a}_{\text{interaction length}} .$$

Thus, the typical wave functions of a particle is necessarily spread out over a region much bigger than the mean particle distance. The particles hence completely lose their individuality, and behave very quantum (i.e., non-classical) in this sense. Fermions, on the other hand, behave much more classical, since for them $\ell \sim \rho^{-1/3}$.

The proof of the lower bound on $E_0(N)$ contains two main steps. First, one would like to replace the hard interaction potential v by a soft one, at the expense of kinetic energy. This softer interaction will have a range R, with $a \ll R \ll \rho^{-1/3}$. Specifically, consider $x_2,\ldots,x_N$ to be fixed for the moment, and assume also that $|x_j - x_k| \geq 2R$ for all $j,k \geq 2$. That is, assume that the balls $B_R(x_j)$ of radius R centered at x_j are non-overlapping. Then

$$\begin{aligned}
&\int \left(|\nabla_1\psi|^2 + \tfrac{1}{2}\sum_{j\geq 2} v(|x_1 - x_j|)|\psi|^2 \right) dx_1 \\
&\geq \sum_{j\geq 2} \int_{B_R(x_j)} \left(|\nabla_1\psi|^2 + \tfrac{1}{2}v(|x_1 - x_j|)|\psi|^2\right) dx_1 \\
&\geq \sum_{j\geq 2} \int U_R(x_1 - x_j)|\psi|^2 dx_1,
\end{aligned} \tag{2.9}$$

where

$$U_R(x) = \begin{cases} e(a,R) & |x| \leq R \\ 0 & |x| > R \end{cases}$$

and $e(a,R)$ is the lowest eigenvalue of $-\Delta + \frac{1}{2}v$ on B_R, with Neumann boundary conditions. As we have already argued in Sect. 2.4.2, the latter is easily seen to be equal to

$$e(a,R) \approx \frac{4\pi a}{|B_R|} \quad \text{for } a \ll R, \tag{2.10}$$

where $|B_R| = (4\pi/3)R^3$ is the volume of B_R.

This is the desired replacing of v by the soft potential U_R. Repeating the above argument for all other particles, we conclude that

$$H_N \geq \sum_{i \neq j} U_R(x_i - x_j)\chi \tag{2.11}$$

where χ is a characteristic function that makes sure that the balls above do not intersect. That is, χ removes three-body collisions. In other words, when three particles come close together, we just drop part of the interaction energy. Since $v \geq 0$, this is legitimate for a lower bound.

The soft potential U_R now predicts the correct energy in first order perturbation theory. In fact, for a constant wave function, the expected value of the right side of (2.11) is approximately equal to $4\pi a N^2/L^3$, with small corrections coming from χ, the region close to the boundary of the box $[0,L]^3$, as well as the fact that (2.10) is only valid approximately.

To make this perturbative argument rigorous, one keeps a bit of the kinetic energy, and uses

$$H_N \geq -\epsilon \sum_{i=1}^{N} \Delta_i + (1-\epsilon) \sum_{i \neq j} U_R(x_i - x_j)\chi \tag{2.12}$$

(using positivity of v). First order perturbation theory can easily be seen to be correct if the perturbation is small compared to the gap above the ground state energy of the unperturbed operator. The gap in the spectrum of $-\epsilon \sum_{i=1}^{N} \Delta_i$ above zero is of the order ϵ/L^2, which has to be compared with $aN\rho$. That is, if

$$L^3 a\rho^2 \ll \frac{\epsilon}{L^2}, \quad \text{or } L^5 \ll \frac{\epsilon}{a\rho^2} \tag{2.13}$$

then the second term in (2.12) is truly a small perturbation to the first term and first order perturbation theory can be shown to yield the correct result for the ground state energy.[2]

[2]Strictly speaking, it is not the expectation value of the perturbation that is the relevant measure of its smallness, but rather the variance. Hence the condition for validity of perturbation theory is slightly more stringent than what is displayed here.

Condition (2.13) is certainly not valid in the thermodynamic limit. To get around this problem, one divides the large cube $[0, L]^3$ into many small cubes of side length ℓ, with ℓ satisfying

$$\ell^5 \ll \frac{\epsilon}{a\rho^2} = \rho^{-5/3} \frac{\epsilon}{a\rho^{1/3}}$$

For an appropriate choice of ϵ, the last fraction is big, hence ℓ can be chosen much larger than the mean particle spacing $\rho^{-1/3}$.

Dividing up space and distributing particles optimally over the cells gives a lower bound to the energy, due to the introduction of additional Neumann boundary conditions on the boundary of the cells. I.e.,

$$E_0(N, L) \geq \min_{\{n_i\}} \sum_i E_0(n_i, \ell)$$

where the minimum is over all distribution of the $N = \sum_i n_i$ particles over the small boxes. Since the interaction is repulsive, it is best to distribute the particles uniformly over the boxes. Hence

$$E_0(N, L) \geq \left(\frac{\ell}{L}\right)^3 E_0(\rho\ell^3, \ell)$$

For our choice of ℓ, we have $E_0(\rho\ell^3, \ell) \approx 4\pi a\rho^2\ell^3$, as explained above, and hence

$$E_0(N, L) \approx \left(\frac{\ell}{L}\right)^3 4\pi a\rho^2\ell^3 = 4\pi a N\rho$$

This concludes the proof, or at least the sketch of the main ideas.

2.4.4 Further Rigorous Results

Extending the method presented in the previous subsection, further results about the low-density behavior of quantum gases have been proved. These include

- *Two-Dimensional Bose Gas.* For a Bose gas in two spatial dimensions, it turns out that [21]

$$e_0(\rho) = \frac{4\pi\rho}{|\ln(a^2\rho)|} \quad \text{for } a^2\rho \ll 1$$

  An interesting feature of this formula is that it does *not* satisfy $E_0(N) \approx \frac{1}{2}N(N-1)E_0(2)$, as it does in three dimensions. The reason for the appearance of the logarithm is the fact that the solution of the zero energy scattering equation

$$-\Delta\phi(x) + \tfrac{1}{2}v(|x|)\phi(x) = 0$$

in two dimensions does not converge to a constant as $|x| \to \infty$, but rather goes like $\ln(|x|/a)$, with a the scattering length.

- *Dilute Fermi Gases.* For a (three-dimensional) Fermi gas at low density ρ, one has [23]

$$e_0(\rho) = \frac{3}{5}\left(\frac{6\pi^2}{q}\right)^{2/3} \rho^{2/3} + 4\pi a \left(1 - \frac{1}{q}\right) \rho + \text{higher order in } \rho$$

Here, q is the number of spin states, i.e., the fermions are considered to have spin $\frac{1}{2}(q-1)$. The first term is just the ground state energy of an ideal Fermi gas. The leading order correction due to the interaction is the same as for bosons, except for the presence of the additional factor $(1 - q^{-1})$. Its presence is due to the fact that the interaction between fermions in the same spin states is suppressed, since for them the spatial part of the wave function is antisymmetric and hence vanishes when the particles are at the same location.

A similar result can also be obtained for a two-dimensional Fermi gas [23].

- *Bose Gas at Positive Temperature.* For a dilute Bose gas at positive temperature $T = 1/\beta$, the natural quantity to investigate is the free energy. For an ideal Bose gas, the free energy per unit volume is given by (2.3), and we shall denote this expression by $f_0(\beta, \rho)$. For an interacting gas, one has

$$f(\beta, \rho) = \underbrace{f_0(\beta, \rho)}_{\text{ideal gas}} + 4\pi a \left(2\rho^2 - [\rho - \rho_c(\beta)]_+^2\right) + \text{higher order} \tag{2.14}$$

and this formula is valid for $a^3\rho \ll 1$ but $\beta\rho^{2/3} \gtrsim O(1)$. Here, $\rho_c(\beta)$ is the critical density for BEC of the ideal gas, given in (2.2), and $[\cdot]_+$ denotes the positive part. That is, $[\rho - \rho_c(\beta)]_+$ is nothing but the condensate density (of the ideal gas).

Since $\rho_c(\beta) \to 0$ as $\beta \to \infty$, (2.14) reproduces the ground state energy formula $4\pi a\rho^2$ in the zero-temperature limit. Above the critical temperature, i.e., for $\rho < \rho_c(\beta)$, the leading order correction is $8\pi a\rho^2$ instead of $4\pi a\rho^2$. The additional factor 2 can be understood as arising from the symmetry requirement on the wavefunctions. Because of symmetrization, the probability that two bosons are at the same locations is twice as big than on average. This applies only to bosons in different modes, however, since if they are in the same mode, symmetrization has no effect. Hence the subtraction of the square of the condensate density, which does not contribute to the factor 2.

The lower bound on $f(\beta, \rho)$ of the form (2.14) was proved in [29], and the corresponding upper bound in [35]. Both articles are rather lengthy and involved, and there are lots of technicalities to turn the above simple heuristics into rigorous bounds. A corresponding result can also be obtained for fermions, as was shown in [28].

For further results and more details, we refer the interested reader to [24].

2.4.5 The Next Order Term

One of the main open problems concerning the ground state energy of a dilute Bose gas concerns the next order term in an expansion for small ρ. It is predicted to equal

$$e_0(\rho) = 4\pi a\rho\left(1 + \frac{128}{15\sqrt{\pi}}\left(a^3\rho\right)^{1/2} + \text{higher order}\right) \tag{2.15}$$

This formula was first derived by Lee–Huang–Yang [13, 14], but is essentially already contained in Bogoliubov's famous 1947 paper [2]. The correction term in (2.15) does not have a simple heuristic explanation, but is a truly quantum-mechanical many-body correlation effect.

The way Bogoliubov arrived at this prediction is the following. The starting point is the Hamiltonian on Fock space. We use plane waves as a basis set, and assume periodic boundary conditions. Then

$$H = \sum_p p^2 a_p^\dagger a_p + \frac{1}{2V}\sum_{p,r,s}\widehat{v}(p) a_{s+p}^\dagger a_{r-p}^\dagger a_r a_s$$

where $V = L^3$ is the volume and

$$\widehat{v}(p) = \int_{\mathbb{R}^3} v(|x|)e^{-ip\cdot x}dx$$

denotes the Fourier transform of v. All sums are over $(\frac{2\pi}{L}\mathbb{Z})^3$. Bogoliubov introduced two approximations, based on the assumption that in the ground state most particles occupy the zero momentum mode. For this reason, one first neglects all terms in H that are higher than quadratic in $a_p^\dagger$ and a_p for $p \neq 0$. Second, one replaces all $a_0^\dagger$ and a_0 by a number $\sqrt{N_0}$, since these operators are expected to have macroscopic values, while their commutator is only one and hence negligible.

The resulting expression for H is the Bogoliubov Hamiltonian

$$\begin{aligned} H_B = {} & \frac{N^2 - (N - N_0)^2}{2V}\widehat{v}(0) \\ & + \sum_{p\neq 0}\left[\left(p^2 + \frac{N_0}{V}\widehat{v}(p)\right)a_p^\dagger a_p + \frac{N_0}{2V}\left(a_p^\dagger a_{-p}^\dagger + a_p a_{-p}\right)\right] \end{aligned} \tag{2.16}$$

This Hamiltonian is now quadratic in creation and annihilation operators, and can be diagonalized easily with the help of a Bogoliubov transformation. The resulting expression for the ground state energy per particles in the thermodynamic limit is

$$4\pi\rho(a_0 + a_1) + 4\pi\rho a_0\frac{128}{15\sqrt{\pi}}\left(a_0^3\rho\right)^{1/2} + \text{higher order in } \rho \tag{2.17}$$

where a_0 and a_1 are, respectively, the first and second order Born approximation to the scattering length a. Explicitly,

$$a_0 = \frac{1}{8\pi} \int_{\mathbb{R}^3} v(|x|)dx$$

and

$$a_1 = -\frac{1}{(8\pi)^2} \int_{\mathbb{R}^6} \frac{v(|x|)v(|y|)}{|x-y|} dx\, dy$$

Moreover, in the ground state

$$\left\langle \sum_{p\neq 0} a_p^\dagger a_p \right\rangle \approx N\sqrt{a^3\rho}$$

hence $(N - N_0)^2 \approx N^2 a^3 \rho$ is negligible to the order we are interested in.

The expression (2.17) looks like an expansion of $e_0(\rho)$ simultaneously in small density and weak coupling. It is hence reasonable to expect the validity of (2.15) without the weak coupling assumption. The proof of this fact is still an open problem, however. For smooth interaction potentials, an upper bound of the correct form was recently proved in [34]. There was also some recent progress in [12] concerning the lower bound, where it was shown that Bogoliubov's approximation is correct, as far as the ground state energy is concerned, if one is allowed to rescale the interaction potential v with ρ is a suitable way.

We remark that the Bogoliubov Hamiltonian (2.16) not only gives a prediction about the ground state energy, but also about the excitation spectrum. Diagonalizing H_B leads to an excitation spectrum of the form

$$\sqrt{p^4 + 2p^2\rho\hat{v}(p)}$$

which is linear for small momentum p. The non-zero slope at $p = 0$ is in fact extremely important physically and has many interesting consequences, concerning superfluidity, for instance. It is also confirmed experimentally. A rigorous proof that the Bogoliubov approximation indeed predicts the correct low energy excitation spectrum is still lacking, however. With the notable exception of exactly solvable models in one dimension [5, 11, 16, 17, 32], the only case where Bogoliubov's prediction about the excitation spectrum is rigorously verified is the mean-field or Hartree limit, where the repulsive interaction is very weak and of long range [30].

2.5 Dilute Bose Gases in Traps

2.5.1 The Gross–Pitaevskii Energy Functional

Actual experiments on cold atomic gases concern inhomogeneous systems, since the particles are confined to a trap with soft walls. Let us extend the analysis of the previous sections to see what happens in the inhomogeneous case.

Let $V(x)$ denote the trap potential, and $\rho(x) = |\phi(x)|^2$ the particle density at a point $x \in \mathbb{R}^3$. If V varies slowly, we can use a local density approximation and assume the validity of the formula $4\pi a\rho^2$ for the energy density of a dilute gas even locally. In this way, we arrive at the expression

$$\mathscr{E}^{\mathrm{GP}}(\phi) = \int_{\mathbb{R}^3} |\nabla\phi(x)|^2 dx + \int_{\mathbb{R}^3} V(x)|\phi(x)|^2 dx + 4\pi a \int_{\mathbb{R}^3} |\phi(x)|^4 dx \quad (2.18)$$

which is known as the *Gross–Pitaevskii (GP) functional*. The last two terms are simply the trap energy and the interaction energy density of a dilute gas in the local density approximation. The first gradient term is added to ensure accuracy even at weak or zero interaction. In fact, for an ideal gas, $a = 0$ and hence (2.18) is certainly the correct description of the energy of the system in this case.

Minimizing (2.18) under the normalization constraint $\int |\phi(x)|^2 dx = N$ leads to the GP energy

$$E^{\mathrm{GP}}(N, a) = \min\left\{\mathscr{E}^{\mathrm{GP}}(\phi) : \int |\phi(x)|^2 dx = N\right\} \quad (2.19)$$

Using standard techniques of functional analysis (see Appendix A of [22]) one can show that there is a minimizer for this problem, which is moreover unique up to a constant phase factor. This holds under suitable assumptions on the trap potential $V(x)$, e.g., if V is locally bounded and tends to infinity as $|x| \to \infty$. The minimizer satisfies the corresponding Euler–Lagrange equation

$$-\Delta\phi + V\phi + 8\pi a|\phi|^2\phi = \mu\phi,$$

which is a nonlinear Schrödinger equation called the GP equation. The chemical potential μ equals $\partial E^{\mathrm{GP}}/\partial N$ and is the appropriate Lagrange parameter to take the normalization condition on ϕ into account.

Based on the discussion above, one would expect that

$$E_0 \approx E^{\mathrm{GP}}$$

and also

$$\rho_0(x) \approx |\phi^{\mathrm{GP}}(x)|^2$$

where E_0 and ρ_0 are the ground state energy and corresponding particle density, respectively. This approximation should be valid if V varies slowly and the gas is sufficiently dilute.

Notice that the GP energy $E^{\mathrm{GP}}(N, a)$ and the corresponding minimizer $\phi^{\mathrm{GP}}_{N,a}$ satisfy the simple scaling relations

$$E^{\mathrm{GP}}(N, a) = N E^{\mathrm{GP}}(1, Na) \quad \text{and} \quad \phi^{\mathrm{GP}}_{N,a}(x) = \sqrt{N}\phi^{\mathrm{GP}}_{1,Na}(x)$$

i.e., Na is the only relevant parameter for the GP theory. In particular, for the purpose of deriving the GP theory from the many-body problem, it makes sense to take N large while Na is fixed. The latter quantity should really be thought of as Na/L, where L is the length scale of the trap V. Hence $a/L \sim N^{-1}$, i.e., V varies indeed much slower that the interaction potential. We shall choose units to make $L = 1$, which simplifies the notation.

Since V is now fixed, we have to rescale the interaction potential v. The appropriate way to do this is to write

$$v_a(|x|) = \frac{1}{a^2} w(|x|/a)$$

for some fixed w. It is then easy to see that v_a has scattering length a if w has scattering length 1. The appropriate many-body Hamiltonian under consideration is thus

$$H_N = \sum_{i=1}^{N} (-\Delta_i + V(x_i)) + \sum_{i<j} v_a(|x_i - x_j|).$$

In this way, a enters as a parameter which can now be varied with N. In particular, the ground state energy $E_0 = \inf \operatorname{spec} H_N$ is now a function of N and a. We shall therefore write $E_0(N, a)$, but suppress the dependence on N and a of the ground state density $\rho_0(x)$ in the notation for simplicity.

Theorem 3 (Lieb–Seiringer–Yngvason [22])

$$\lim_{N\to\infty} \frac{E_0(N, g/N)}{N} = E^{\mathrm{GP}}(1, g) \quad \textit{for any } g \geq 0$$

In the same limit

$$\lim_{N\to\infty} \frac{1}{N} \rho_0(x) = \left| \phi_{1,g}^{\mathrm{GP}}(x) \right|^2$$

Note that in the limit under consideration $a^3 \bar{\rho} \sim N^{-2}$, where $\bar{\rho} \sim N$ denotes the average density. In particular, the gas is very dilute if $g = Na$ is fixed. The result of Theorem 3 is actually uniform in g as long $a^3 \bar{\rho} \to 0$ as $N \to \infty$. I.e., g is allowed to go to ∞ with N at a suitable rate, as long as the gas stays dilute.

The proof of Theorem 3 is similar to the homogeneous case, and uses the same ideas in the lower bound. In particular, space is divided up into small boxes and the particles are then distributed optimally over these boxes. In the inhomogeneous case considered here, the distribution will be non-uniform, of course.

2.5.2 BEC of Dilute Trapped Gases

So far the discussion has focused on the ground state energy and the corresponding particle density. But what about BEC? As discussed in Sect. 2.3, BEC is a property of the reduced one-particle density matrix of the system. Specifically, if Ψ_0 is the ground state of H_N, then the one-particle density matrix γ is the operator on $L^2(\mathbb{R}^3)$ with integral kernel

$$\gamma_0(x, x') = N \int \Psi_0(x, x_2, \ldots, x_N)\Psi_0(x', x_2, \ldots, x_N)^* dx_2 \cdots dx_N$$

Recall that γ is a positive trace-class operator, with $\operatorname{Tr}\gamma = N$.

Theorem 4 (Lieb–Seiringer [18]) *In the same limit as in Theorem 3*

$$\lim_{N\to\infty} \frac{1}{N}\gamma_0 = \left|\phi_{1,g}^{\mathrm{GP}}\right\rangle\left\langle\phi_{1,g}^{\mathrm{GP}}\right|$$

What the theorem says is that there is complete BEC, in the sense that the largest eigenvalue of γ_0, divided by N, is not only non-zero but actually equal to one in the dilute limit considered. I.e., the one-particle density matrix becomes a rank-one projection in the limit, just like for a non-interacting gas. The condensate wave function $\phi_{1,g}^{\mathrm{GP}}$ still depends on the interacting strength via g, however, and might have very little overlap with the non-interacting state at $g = 0$ if g is large.

Theorem 4 represents the only known case of a continuous system with genuine interactions where BEC has been proved. The proof is so far restricted to zero temperature and to the very dilute limit where $a^3\bar{\rho} \sim N^{-2}$ as $N \to \infty$.

Before discussing the proof of Theorem 4, we shall first generalize the setting in a non-trivial way by allowing the system to rotate about a fixed axis. Theorem 4 can thus be considered as a special case of a more general result to be discussed next.

2.5.3 Rotating Bose Gases

An interesting property of dilute cold Bose gases is their response to rotation. In fact, rotating Bose–Einstein condensates are nowadays routinely created in the lab, by stirring the system much like coffee with a spoon.

Even though the system under consideration is now rotating, we can still think of it as being at equilibrium if we go to the rotating frame of reference. Just like in classical mechanics, the only effect of this transformation on the Hamiltonian is to add a term proportional to the total angular momentum. More precisely,

$$H_N \longrightarrow H_N - \Omega \cdot L$$

where $\Omega \in \mathbb{R}^3$ denotes the angular velocity (having an axis and a magnitude) and $L = \sum_{i=1}^N L_i$ denotes the total angular momentum of the system.

In the experiments on rotating gases, one observes the appearance of quantized vortices, related to the superfluid properties of the system. This is schematically sketched in Fig. 2.3.

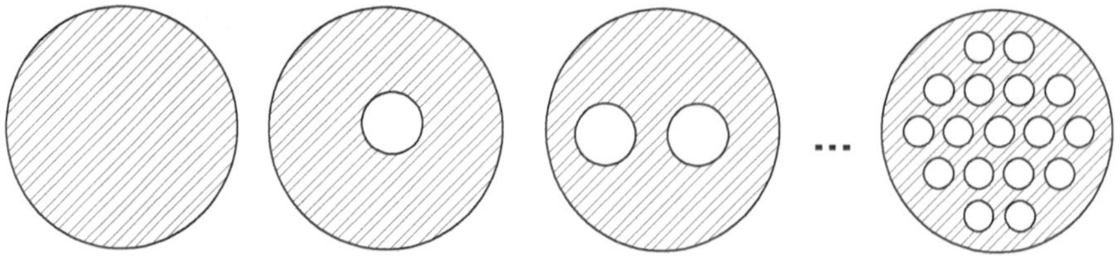

Fig. 2.3 Quantized vortices in a rotating Bose condensate, showing up as holes in the density. More and more vortices appear as the angular velocity is increased. For actual snapshots of experiments, see http://jila.colorado.edu/bec/hi_res_pic_album_macromedia or http://www.bec.nist.gov/gallery.html

The quantized vortices can also be seen by minimizing the appropriate GP functional, which now reads

$$\mathscr{E}^{\rm GP}(\phi) = \langle \phi \,|-\Delta + V(x) - \Omega \cdot L|\, \phi\rangle + 4\pi a \int_{\mathbb{R}^3} |\phi(x)|^4 dx$$

with $L = -ix \wedge \nabla$, as usual. In order for the confining force to overcome the centrifugal force, we have to assume that

$$V(x) - \frac{1}{4}|\Omega \wedge x|^2 \tag{2.20}$$

is bounded below and goes to infinity at infinity. Under this condition, one can still prove the existence of a minimizer of the GP functional. In general it will not be unique anymore, however. This non-uniqueness is related to spontaneous symmetry breaking. In fact, if V is rotation symmetric about the Ω axis, i.e, $[V, \Omega \cdot L] = 0$, then $\mathscr{E}^{\rm GP}$ is invariant under rotations about this axis. In general, a minimizer $\phi^{\rm GP}$ will not have this symmetry, however, due to the appearance of quantized vortices. If there are more than one, these obviously can not be arranged in a symmetric way. I.e, in general we expect a whole continuum of minimizers in the case the GP functional is rotation symmetric.

The N-body Hamiltonian under consideration now is

$$H_N = \sum_{i=1}^{N}(-\Delta_i + V(x_i) - \Omega \cdot L_i) + \sum_{i<j} v_a(|x_i - x_j|)\,.$$

Its ground state energy will be denoted by $E_0(N, a, \Omega)$.

Theorem 5 (Lieb–Seiringer [19]) *For any $g \geq 0$ and $\Omega \in \mathbb{R}^3$ (subject to the constraint that (2.20) is bounded from below and goes to infinity at infinity)*

$$\lim_{N\to\infty} \frac{E_0(N, g/N, \Omega)}{N} = E^{\rm GP}(1, g, \Omega)$$

Moreover, up to a subsequence, the one-particle density matrix of a ground state (or any approximate ground state, in fact) satisfies

$$\lim_{N\to\infty} \frac{1}{N}\gamma = \int d\mu \, |\phi^{\mathrm{GP}}\rangle\langle\phi^{\mathrm{GP}}| \tag{2.21}$$

where $d\mu$ is a probability measure on the set of minimizers of $\mathcal{E}^{\mathrm{GP}}$.

By an approximate minimizer we mean a state that has the same energy as the ground state energy, to leading order. In other words, a state with energy equal to

$$\lim \frac{1}{N}\langle H_N\rangle = E^{\mathrm{GP}}(1, g, \Omega)$$

in the limit $N \to \infty$, $Na \to g$.

Equation (2.21) is the natural generalization of complete BEC in the case of non-uniqueness of GP minimizers. Because of the linearity of quantum mechanics, the best one can hope for is a convex combination of completely condensed states. In fact, (2.21) can also be seen as establishing the spontaneous symmetry breaking mentioned earlier. Under an infinitesimal perturbation, the GP functional will generically have a unique minimizer, and Theorem 5 in this case implies that there is then complete BEC in the usual sense.

We note that the bosonic symmetry requirement on the N-particle wave functions is crucial for Theorem 5 to hold. In contrast, for the discussion of the ground state of non-rotating systems, Bose symmetry does not have to be enforced explicitly, it comes out automatically as the ground state of an operator of the form $-\Delta + W(x)$ is always unique and positive and hence can only be permutation symmetric. For rotating systems, this is generally not the case, and Bose symmetry can not be ignored.

2.5.4 Main Ideas in the Proof

We split the proof of Theorem 5 into three parts.

Step 1. The first step is again to try to replace the hard interaction potential $v_a(|x|)$ by a softer one, $U_R(|x|)$, at the expense of some kinetic energy. We must not use up all the kinetic energy as we did in the homogeneous case, however, since we still need to obtain the gradient term in the GP functional. The key idea is to split the kinetic energy into a high momentum and a low momentum part. Only the high momentum part $|p| \geq p_c$ will be needed to achieve the replacement $v_a \to U_R$, while the low momentum part $|p| \leq p_c$ will kept as it is needed in the GP functional. We will, in fact choose

$$1 \ll p_c \ll \frac{1}{R} \tag{2.22}$$

The first condition implies that only momenta irrelevant for the GP functional are being used, while the second makes sure that all momenta relevant on the length scale of U_R are actually employed. The crucial Lemma that improves (2.9) is the following. Its proof is in [23].

Lemma 1 *Let $\chi_{B_R(0)}$ denote the characteristic function of the ball of radius R centered at the origin. For any $0 < \epsilon < 1$,*

$$-\nabla \cdot \xi(p)\chi_{B_R(0)}(x)\xi(p)\nabla + \frac{1}{2}v(|x|) \geq (1-\epsilon)U_R(|x|) - \frac{1}{\epsilon}w_R(x) \qquad (2.23)$$

where

$$w_R(x) = \frac{2a}{\pi^2} f_R(x) \int_{\mathbb{R}^3} f_R(y)dy$$

and $f_R(x) = \sup_{|y|\leq R} |h(x-y) - h(x)|$, $\widehat{h}(p) = 1 - \xi(p)$.

The function ξ is chosen to be a smooth characteristic function of the set $\{|p| \geq p_c\}$. Hence the first term in (2.23) is a version of the Laplacian that has been restricted to high momentum and localized to a ball of radius R (centered at the origin). The price one has to pay for the cut-off ξ is the error term w_R, which is supported also outside the ball but can be made to decay very fast by choosing ξ smooth. For our choice of p_c in (2.22), it will be negligible compared to U_R.

Lemma 1 implies the operator lower bound

$$\begin{aligned} H_N \geq & \sum_{i=1}^N \left(-\Delta_i(1-\xi(p_i)^2) + V(x) - \Omega \cdot L_i\right) \\ & + \sum_{i\neq j}\left((1-\epsilon)U_R(|x_i - x_j|) - \frac{1}{\epsilon}w_R(|x_i - x_j|)\right)\chi \end{aligned} \qquad (2.24)$$

where χ is again a characteristic function that excludes three- and more particle collisions.

Step 2. In order to proceed, we want to get rid of both the w_R term and the characteristic function χ in (2.24). For this purpose, we need some a priori bounds that tell us that the expected values of w_R and $1-\chi$ in the ground state of H_N are not too big. For this purpose, we obtained some rough bounds on the three-particle density of a ground state of H_N, using path integrals. These bounds are of the form

$$\left\langle f(x_1, x_2, x_3)\right\rangle \leq \Lambda(\alpha, f)\, e^{\alpha(E_0(N) - E_0(N-3))}$$

where $\langle\cdot\rangle$ denotes expectation in the zero-temperature state, f is an arbitrary positive bounded function, $\alpha > 0$ is arbitrary and $\Lambda(\alpha, f)$ denotes the largest eigenvalue of the operator

$$\sqrt{f}e^{-\alpha(-\Delta_1-\Delta_2-\Delta_3+V(x_1)+V(x_2)+V(x_3))}\sqrt{f}$$

on $L^2(\mathbb{R}^9)$. This bound is certainly not optimal, but suffices to show that the terms in question do not contribute to the ground state energy to the order we are interested. I.e., we conclude that

$$\inf \operatorname{spec} H_N \geq \inf \operatorname{spec} \widetilde{H}_N - \delta N$$

where $\delta \to 0$ in the limit considered. Moreover,

$$\widetilde{H}_N = \sum_{i=1}^{N} \left(-\Delta_i(1-\xi(p_i)^2) + V(x) - \Omega \cdot L_i\right) + \sum_{i \neq j} U_R(|x_i - x_j|) \qquad (2.25)$$

That is, we have managed to genuinely replace the hard interaction potential v_a by the soft one U_R, at the expense of the high-momentum part of the kinetic energy, as well as a minor shift in the ground state energy.

Step 3. Let us denote the one-particle part of the Hamiltonian $\widetilde{H}_N$ by h for simplicity, i.e.,

$$h = -\Delta(1-\xi(p)^2) + V(x) - \Omega \cdot L$$

In second quantized form, using as a basis the eigenstates of h, we have

$$\widetilde{H} = \sum_i \langle \varphi_i | h | \varphi_i \rangle a_i^\dagger a_i + \sum_{ijkl} \langle \varphi_i \otimes \varphi_j | U_R | \varphi_k \otimes \varphi_l \rangle a_i^\dagger a_j^\dagger a_k a_l \qquad (2.26)$$

Notice that if we ignore all commutators between the $a_i^\dagger$ and a_i and treat them as numbers, z_i^* and z_i, respectively, (2.26) becomes

$$\langle \Phi | h | \Phi \rangle + \int_{\mathbb{R}^6} |\Phi(x)|^2 U_R(|x-y|) |\Phi(y)|^2 \, dx \, dy$$

with

$$\Phi(x) = \sum_i z_i \varphi_i(x)$$

This is essentially the GP functional, except for the cutoff in the kinetic energy, which is irrelevant for $p_c \gg 1$, and the fact that the interaction is U_R instead of $4\pi a \delta$. Since $R \ll 1$, however, and $\int U_R = 4\pi a$, U_R is an approximate δ function with the correct coefficient.

In other words, the GP functional emerges from the many-body Hamiltonian on Fock space in a classical limit, replacing all the creation and annihilation operators by complex numbers. In this sense, GP theory is a classical field approximation to the quantum field theory defined by $\widetilde{H}$. Note that this is only true for the low momentum part, however. It is important that we have already completed Step 1 above to replace the true interaction potential v_a by U_R. If we had not done so, the classical approximation would also look like a GP functional, but with the wrong coefficient $\frac{1}{2}\int v$ instead of $4\pi a$ in front of the quartic term.

What remains to be done is to investigate the validity of the replacement of the creation and annihilation operators by numbers. This can be conveniently done using coherent states. We shall describe what these are in the next subsection, and complete the sketch of the proof of Theorem 5 there.

2.5.5 *Coherent States*

With $|0\rangle$ denoting the Fock space vacuum, and $z \in \mathbb{C}$, consider the state on Fock space

$$|z\rangle = e^{za^\dagger - z^* a}|0\rangle$$

where a and $a^\dagger$ are the annihilation and creation operators for one particular mode. Since the exponent is anti-hermitian, $|z\rangle$ is a vector of length one. Because of $[a, a^\dagger] = 1$, one can rewrite it also as

$$|z\rangle = e^{-|z|^2/2} e^{za^\dagger}|0\rangle$$

This state is a superposition of all states with different particle number in the mode under consideration. As z varies over the complex plane $\mathbb{C}$, the states $|z\rangle$ span the whole Fock space associated with the mode a. In fact, one can easily check the completeness relation

$$\int_{\mathbb{C}} \frac{dz}{\pi} |z\rangle\langle z| = 1$$

where dz stands for the standard Lebesgue measure $dx\, dy$, $z = x + iy$. States with different value of z are of course not orthogonal. One can also check that

$$a|z\rangle = z|z\rangle$$

i.e., $|z\rangle$ is an eigenstate of a with eigenvalue z.

For a general operator given in terms of a and $a^\dagger$ (typically a polynomial), define its *lower symbol* $h_l(z)$ by

$$h_l(z) = \langle z|h|z\rangle$$

Note that if h is normal ordered, i.e, all creation operators appear to the left of all annihilation operators, then $h_l(z)$ is obtained from h simply by replacing all a's by z and all $a^\dagger$'s by z^*. Many operators (in particular, all polynomials) also have *upper symbols*, which are functions $h_u(z)$ such that

$$h = \int_{\mathbb{C}} \frac{dz}{\pi} h_u(z)|z\rangle\langle z|$$

In fact, $h_u(z)$ is obtained by replacing a by z and $a^\dagger$ by z^* in the anti-normal ordered form of h.

Examples:

h	$h_l(z)$	$h_u(z)$
a	z	z
$a^\dagger$	z^*	z^*
$a^\dagger a$	$\lvert z\rvert^2$	$\lvert z\rvert^2 - 1$

In general, one can show that $h_l(z)$ and $h_u(z)$ are related by

$$h_u(z) = e^{-\partial_{z^*}\partial_z} h_l(z)$$

(as long as the right side exists).

Note that for self-adjoint h

$$\inf_z h_u(z) \leq \inf \operatorname{spec} h \leq \inf_z h_l(z)$$

The same is true for partition functions, namely the Berezin–Lieb inequalities

$$\int_{\mathbb{C}} \frac{dz}{\pi} e^{-h_l(z)} \leq \operatorname{Tr} e^{-h} \leq \int_{\mathbb{C}} \frac{dz}{\pi} e^{-h_u(z)}$$

hold. These inequalities are, in fact, the origin of the terminology "upper" and "lower" symbols; Upper symbols give upper bounds to the partition function, while lower symbols give lower bounds.

Effectively, coherent states replace a quantum problem by a classical problem with phase space $\mathbb{C}$, replacing creation and annihilation operators by numbers. Note that the difference in the upper and lower bounds comes from the difference in the upper and lower symbols, in particular the factor -1 for the quadratic operator $a^\dagger a$ in the example above.

Coherent states can be used for many modes at the same time, simply using tensor products. One can not use them for *all* modes, however. Even for the number operator, the upper and lower symbols differ by a constant which is the number of modes, and we want to avoid infinities.

Let us split the Fock space into two parts,

$$\mathscr{F} = \mathscr{F}_< \otimes \mathscr{F}_>$$

corresponding to the splitting of the one-particle Hilbert space $\mathscr{H}_1$ into $\mathscr{H}_< \oplus \mathscr{H}_>$, where $\mathscr{H}_<$ is a finite dimensional space spanned by the modes $\varphi_1, \ldots, \varphi_J$. Here, $J \geq 0$ is some large finite number to be determined later. On $\mathscr{F}_<$, we can use coherent states for all the modes. In particular, for $\widetilde{H}$, our Hamiltonian under consideration, we can write

$$\widetilde{H} = \int_{\mathbb{C}^J} \prod_{j=1}^{J} \frac{dz_j}{\pi} |z_1 \otimes \cdots \otimes z_J\rangle\langle z_1 \otimes \cdots \otimes z_J| \otimes K(z_1, \ldots, z_J)$$

where the upper symbol $K(z_1, \ldots, z_J)$ is now an operator on $\mathscr{F}_>$, the Fock space for the large modes. The key point of this decomposition is that

$$\inf \operatorname{spec} \widetilde{H} \geq \inf_{z_1,\ldots,z_J} \inf \operatorname{spec} K(z_1, \ldots, z_J)$$

One can show that, for $J \gg 1$ appropriately chosen

$$K(z_1, \ldots, z_J) = \mathscr{E}^{\mathrm{GP}}(\Phi) + \text{error terms}$$

where

$$\Phi(x) = \sum_{j=1}^{J} z_j \varphi_j(x)$$

The error terms are still operators on $\mathscr{F}_>$, but they are small (at least in expectation) for an appropriate choice of the interaction range R. For these estimates, it is important that $R \gg a \sim N^{-1}$, hence the necessity of Step 1. If fact, the larger R the better the control of the error terms, but we can still get away with some $R \ll N^{-1/3}$, as required.

This completes the sketch of the proof of the lower bound to the ground state energy in Theorem 5. For the details, we refer to [19]. An appropriate upper bound can be derived using the variational principle [27].

For the proof of BEC, one can proceed as above, but adding to the one-particle Hamiltonian some perturbation S. The proof goes through essentially without change, since the precise form of h has never been used. The result is the validity of the GP theory for the ground state energy, even with h replaced by $h + S$. One can now use standard convexity theory, differentiating with respect to S. The key point is this: If concave functions $f_n(x)$ converge pointwise to a function f, then the right and left derivatives f'_+ and f'_- (which always exist for concave functions) satisfy

$$f'_+(x) \leq \liminf_{n\to\infty} f'_{n,+}(x) \leq \limsup_{n\to\infty} f'_{n,-}(x) \leq f'_-(x) \tag{2.27}$$

In particular, if f is differentiable at a point x, then there is equality everywhere in (2.27).

The left and right derivatives of

$$\lambda \mapsto \inf_{\phi} \left(\mathscr{E}^{\mathrm{GP}}(\phi) + \lambda\langle\phi|S|\phi\rangle \right)$$

at $\lambda = 0$ are both of the form $\langle\phi^{\mathrm{GP}}|S|\phi^{\mathrm{GP}}\rangle$, with ϕ^{GP} a GP minimizer (in the case $\lambda = 0$). They need not be the same, however. We thus conclude that

$$\min_{\phi^{\mathrm{GP}}} \langle \phi^{\mathrm{GP}} | S | \phi^{\mathrm{GP}} \rangle \leq \lim_{N \to \infty} \frac{1}{N} \operatorname{Tr} S\gamma \leq \max_{\phi^{\mathrm{GP}}} \langle \phi^{\mathrm{GP}} | S | \phi^{\mathrm{GP}} \rangle \tag{2.28}$$

for the one-particle density matrix γ of a ground state of H_N, where the maximum and minimum, respectively, is over all minimizers of the GP functional. Since (2.28) is valid for all (hermitian, bounded) S, the statement about BEC follows now quite easily. For simplicity, just consider the case of a unique GP minimizer, in which case there is equality in (2.28). It is easy to see that this implies $\lim_{N\to\infty} N^{-1}\gamma = |\phi^{\mathrm{GP}}\rangle\langle\phi^{\mathrm{GP}}|$. The more general case is discussed in detail in [19].

2.5.6 Rapidly Rotating Bose Gases

Consider now the special case of a harmonic trap potential

$$V(x) = \frac{1}{4}|x|^2$$

This is of particular relevance for the experimental situation, where the trap potential is typically close to being harmonic. The one-particle part of the Hamiltonian can then be written as

$$h = -\Delta + V(x) - \Omega \cdot L = \left(-i\nabla - \tfrac{1}{2}\Omega \wedge x\right)^2 + \tfrac{1}{4}\left(|x|^2 - |\Omega \wedge x|^2\right)$$

The first part on the right side is the same as the kinetic energy of a particle in a homogeneous magnetic field Ω and, in particular, is translation invariant (up to a gauge transformation). It follows that h is bounded from below only for $|\Omega| \leq 1$. The angular velocity has to be less than one, otherwise the trapping force is not strong enough to compensate the centrifugal force and the system flies apart.

What happens to a dilute Bose gas as $|\Omega|$ approaches 1? For $e = \Omega/|\Omega|$ the rotation axis, let us rewrite h as

$$h = \underbrace{\left(-i\nabla - \tfrac{1}{2}e \wedge x\right)^2 + \tfrac{1}{4}|e \cdot x|^2}_{k} + (e - \Omega) \cdot L$$

For Ω close to e, the last term can be considered as a perturbation of the rest, which we denote by k. The spectrum of k equals $\{3/2, 5/2, 7/2, \ldots\}$, and each energy level is infinitely degenerate. These energy levels are in fact just the Landau levels for a particle in a homogeneous magnetic field (for the motion perpendicular to Ω), combined with a simple harmonic oscillator in the Ω direction.

For $|e - \Omega| \ll 1$, we can thus restrict the one-particle Hilbert space to the lowest Landau level (LLL) when investigating the low energy behavior of the system. This LLL consists of functions of the form

$$f(z)e^{-|x|^2/4}$$

where we use a complex variable z for the coordinates perpendicular to Ω. In particular, $|x|^2 = |z|^2 + |e \cdot x|^2$. Moreover, the function f has to be analytic, i.e., it is an entire function of z. The only freedom lies in the choice of f, in fact, the Gaussian factor is fixed. In particular, the motion in the Ω direction is frozen into the ground state of the harmonic oscillator. Because of that, it is convenient to think of the Hilbert space as the space of analytic functions f only, and absorb the Gaussian into the measure. The resulting space is known as the Bargmann space

$$\mathscr{B} = \left\{ f : \mathbb{C} \to \mathbb{C} \text{ analytic}, \int_{\mathbb{C}} |f(z)|^2 e^{-|z|^2/2} dz < \infty \right\}$$

On the space $\mathscr{B}$, the angular momentum $e \cdot L$ simply acts as

$$L = z \frac{\partial}{\partial z}$$

In particular, its eigenstates are z^n, $n = 0, 1, 2, \ldots$. These states form an orthonormal bases of $\mathscr{B}$. In particular, note that $L \geq 0$ on $\mathscr{B}$.

Having identified the relevant one-particle Hilbert space for the low energy physics of a rapidly rotating Bose gas, what should the relevant many-body Hamiltonian look like? The only term left from the one-particle part of H_N is $(e - \Omega) \cdot L$. If we assume that the interaction is short range, i.e, $a \ll 1$ (where 1 is the relevant "magnetic length" in our units), it can be approximated by a δ-function in the LLL. I.e., we introduce as a many-body Hamiltonian on $\mathscr{B}^{\otimes N}$

$$H_N^{\mathrm{LLL}} = \omega \sum_{i=1}^{N} z_i \frac{\partial}{\partial z_i} + 8\pi a \sum_{1 \leq i < j \leq N} \delta_{ij} \tag{2.29}$$

where $\omega > 0$ is short for $1 - |\Omega|$ and δ_{ij} is obtained from projecting $\delta(x_i - x_j)$ onto the LLL level. Explicitly, we have

$$\left(\delta_{12} f\right)(z_1, z_2) = (2\pi)^{-3/2} f\left(\tfrac{1}{2}(z_1 + z_2), \tfrac{1}{2}(z_1 + z_2)\right)$$

That is, δ_{12} symmetrizes the arguments z_1 and z_2. In particular, it takes analytic functions into analytic functions. Except for the unimportant prefactor $(2\pi)^{-3/2}$, δ_{12} is, in fact, a projection, projecting onto relative angular momentum zero. The factor $8\pi a$ in front of the interaction term in (2.29) is chosen as to reproduce the correct expression for the ground state energy of a homogeneous system.

The introduction of the effective many-body Hamiltonian H_N^{LLL} in the lowest Landau level raises interesting questions. First of all, can one rigorously justify the approximations leading to H_N^{LLL}? In other words, can it be rigorously derived from the full many-body problem, defined by H_N on the entire Hilbert space? This was indeed achieved in [15], where it was shown that for small ω and small a, the low energy spectrum and corresponding eigenstates of H_N are indeed well approximated by the corresponding ones of H_N^{LLL}, and converge to these in the limit $\omega \to 0$, $a \to 0$ with a/ω fixed. Note that H_N^{LLL} can *not* be obtained by simply projecting H_N onto the LLL, as this would not reproduce the correct prefactor $8\pi a$ in front of the interaction. It is important to first integrate out the high energy degrees of freedom, associated with length scales much smaller than 1, as we have done several times earlier. The projection onto the LLL is only a good approximation for length scales of order one and larger.

Having rigorously derived H_N^{LLL} from the full many-body problem, what have we learned? It still remains to investigate the relevant properties of this effective model, in particular its spectrum and corresponding eigenstates. Relatively little is known about these questions, however, despite the apparent simplicity of the model. Interesting behavior reminiscent of the fractional quantum Hall effect in fermionic systems is expected.

Note that H_N^{LLL} is the sum of two terms

$$H_N^{\mathrm{LLL}} = \omega \underbrace{\sum_{i=1}^{N} z_i \frac{\partial}{\partial z_i}}_{L_N} + 8\pi a \underbrace{\sum_{1\le i<j\le N} \delta_{ij}}_{\Delta_N} \tag{2.30}$$

that commute with each other, i.e., $[L_N, \Delta_N] = 0$. It therefore makes sense to look at their joint spectrum. Of particular relevance is the so-called "yrast curve", which is the lowest energy of Δ_N in a given sector of angular momentum. That is,

$$\Delta_N(L) = \inf \operatorname{spec} \Delta_N \restriction_{L_N = L}$$

It is known explicitly for small and large L. In fact, the known values of $\Delta_N(L)$ are

$$\Delta_N(L) = \frac{1}{2(2\pi)^{3/2}} \times \begin{cases} N(N-1) & L \in \{0,1\} \\ N(N-1-\frac{1}{2}L) & 2 \le L \le N \\ 0 & L \ge N(N-1) \end{cases}$$

The minimizer of $L = N(N-1)$ is in fact the bosonic analogue of the Laughlin state

$$\prod_{i<j} (z_i - z_j)^2$$

for which obviously the interaction energy is zero. Note the exponent 2, which has to be even since we are dealing with bosons.

A sketch of the joint spectrum of L_N and Δ_N is given in Fig. 2.4. The interesting part concerns angular momenta of order N^2, in which case Δ_N is of order N. For $L \ll N^2$, one can show that the GP approximation becomes exact. I.e., for large N and $L \ll N^2$, the convex hull of $\Delta_N(L)$ is given by

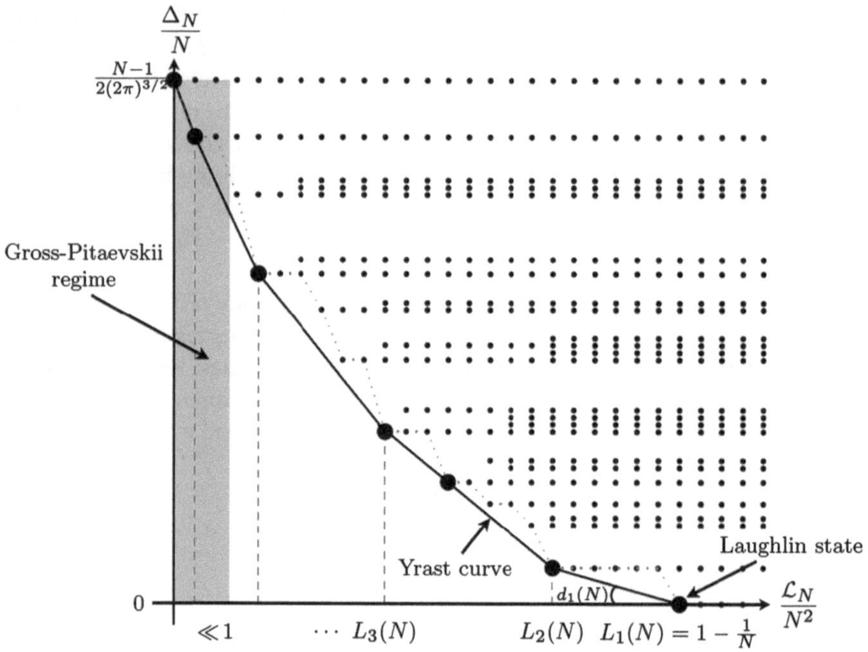

Fig. 2.4 Sketch of the joint spectrum of L_N and Δ_N. The GP approximation is valid in the *shaded region* on the *left*. The *small dots* show the yrast *curve*, in *black* is its convex hull. The *bold dots* on the convex hull are, in fact, the possible ground states of H_N^{LLL} as one varies ω/a

$$\inf_{\phi \in B} \left\{ \int_{\mathbb{C}} |\phi(z)|^4 e^{-|z|^2} dz : \|\phi\|^2 = N, \langle \phi | L | \phi \rangle = L \right\}$$

This was proved in [25] using coherent states. The condition $L \ll N^2$ corresponds to the case when the number of particles is much larger than the number of vortices. Once these two numbers becomes comparable, the GP approximation breaks down and interesting new physics with highly correlated many-body states is expected to occur.

No rigorous bounds on $\Delta_N(L)$ are available for $L \sim N^2$ (but $L < N(N-1)$, of course). In fact, even to prove the existence of the limit

$$\lim_{N \to \infty} \frac{\Delta_N(\ell N^2)}{N}$$

is an open problem. Besides $\Delta_N(L)$, one would also like to understand the existence or non-existence of spectral gaps above the ground state energy (uniformly in the particle number), and other quantities of this type. A lot remains to be done.

Acknowledgements Many thanks to Dana Mendelson, Alex Tomberg and Daniel Ueltschi for allowing me to use their figures in these notes. Financial support through the ERC starting grant CoMboS-239694 is gratefully acknowledged.

References

1. M.H. Anderson, J.R. Ensher, M.R. Matthews, C.E. Wieman, E.A. Cornell, Observation of Bose-Einstein condensation in a dilute atomic vapor. Science **269**, 198–201 (1995)
2. N.N. Bogoliubov, On the theory of superfluidity. J. Phys. (USSR) **11**, 23 (1947)
3. S.N. Bose, Plancks Gesetz und Lichtquantenhypothese. Z. Phys. **26**, 178–181 (1924)
4. E. Buffet, J.V. Pulè, Fluctuation properties of the imperfect Bose gas. J. Math. Phys. **24**, 1608–1616 (1983)
5. F. Calogero, Ground state of a one-dimensional N-body system. J. Math. Phys. **10**, 2197–2200 (1969); Solution of the one-dimensional N-body problems with quadratic and/or inversely quadratic pair potentials. J. Math. Phys. **12**, 419–436 (1971)
6. K.B. Davis, M.O. Mewes, M.R. Andrews, N.J. van Druten, D.S. Durfee, D.M. Kurn, W. Ketterle, Bose-Einstein condensation in a gas of sodium atoms. Phys. Rev. Lett. **75**, 3969–3973 (1995)
7. F.J. Dyson, Ground-state energy of a hard-sphere gas. Phys. Rev. **106**, 20–26 (1957)
8. F.J. Dyson, E.H. Lieb, B. Simon, Phase transitions in quantum spin systems with isotropic and nonisotropic interactions. J. Stat. Phys. **18**, 335–383 (1978)
9. A. Einstein, Quantentheorie des einatomigen idealen Gases. Sitzungsber. Preuss. Akad. Wiss., Phys.-math. Klasse, 261–267 (1924). Zweite Abhandlung. Sitzungsber. Preuss. Akad. Wiss., Phys.-math. Klasse, 3–14 (1925)
10. J. Fröhlich, B. Simon, T. Spencer, Infrared bounds, phase transitions and continuous symmetry breaking. Comm. Math. Phys. **50**, 79–85 (1976)
11. M. Girardeau, Relationship between systems of impenetrable bosons and fermions in one dimension. J. Math. Phys. **1**, 516–523 (1960)
12. A. Giuliani, R. Seiringer, The ground state energy of the weakly interacting Bose gas at high density. J. Stat. Phys. **135**, 915–934 (2009)
13. T.D. Lee, C.N. Yang, Many body problem in quantum mechanics and quantum statistical mechanics. Phys. Rev. **105**, 1119–1120 (1957)
14. T.D. Lee, K. Huang, C.N. Yang, Eigenvalues and eigenfunctions of a Bose system of hard spheres and its low-temperature properties. Phys. Rev. **106**, 1135–1145 (1957)
15. M. Lewin, R. Seiringer, Strongly correlated phases in rapidly rotating Bose gases. J. Stat. Phys. **137**, 1040–1062 (2009)
16. E.H. Lieb, Exact analysis of an interacting Bose gas, II. The excitation spectrum. Phys. Rev. **130**, 1616–1624 (1963)
17. E.H. Lieb, W. Liniger, Exact Analysis of an interacting Bose gas, I. The general solution and the ground state. Phys. Rev. **130**, 1605–1616 (1963)
18. E.H. Lieb, R. Seiringer, Proof of Bose-Einstein condensation for dilute trapped gases. Phys. Rev. Lett. **88**, 170409 (2002)
19. E.H. Lieb, R. Seiringer, Derivation of the Gross-Pitaevskii equation for rotating Bose gases. Comm. Math. Phys. **264**, 505–537 (2006)
20. E.H. Lieb, J. Yngvason, Ground state energy of the low density Bose gas. Phys. Rev. Lett. **80**, 2504–2507 (1998)
21. E.H. Lieb, J. Yngvason, The ground state energy of a dilute two-dimensional Bose gas. J. Stat. Phys. **103**, 509 (2001)
22. E.H. Lieb, R. Seiringer, J. Yngvason, Bosons in a trap: A rigorous derivation of the Gross-Pitaevskii energy functional. Phys. Rev. A **61**, 043602 (2000)

23. E.H. Lieb, R. Seiringer, J.P. Solovej, Ground-state energy of the low-density Fermi gas. Phys. Rev. A **71**, 053605 (2005)
24. E.H. Lieb, R. Seiringer, J.P. Solovej, J. Yngvason, *The Mathematics of the Bose Gas and its Condensation*. Oberwolfach Seminars, vol. 34 (Birkhäuser, Boston, 2005). Also available at arXiv:cond-mat/0610117
25. E.H. Lieb, R. Seiringer, J. Yngvason, Yrast line of a rapidly rotating Bose gas: Gross-Pitaevskii regime. Phys. Rev. A **79**, 063626 (2009)
26. O. Penrose, L. Onsager, Bose-Einstein condensation and liquid helium. Phys. Rev. **104**, 576–584 (1956)
27. R. Seiringer, Ground state asymptotics of a dilute, rotating gas. J. Phys. A Math. Gen. **36**, 9755–9778 (2003)
28. R. Seiringer, The thermodynamic pressure of a dilute Fermi gas. Comm. Math. Phys. **261**, 729–758 (2006)
29. R. Seiringer, Free energy of a dilute Bose gas: Lower bound. Comm. Math. Phys. **279**, 595–636 (2008)
30. R. Seiringer, *The Excitation Spectrum for Weakly Interacting Bosons*. Commun. Math. Phys. **306**, 565–578 (2011)
31. R. Seiringer, D. Ueltschi, Rigorous upper bound on the critical temperature of dilute Bose gases. Phys. Rev. B **80**, 014502 (2009)
32. B. Sutherland, Quantum many-body problem in one dimension: Ground state. J. Math. Phys. **12**, 246–250 (1971); Quantum many-body problem in one dimension: Thermodynamics. J. Math. Phys. **12**, 251–256 (1971)
33. D. Ueltschi, Feynman cycles in the Bose gas. J. Math. Phys. **47**, 123303 (2006)
34. H.-T. Yau, J. Yin, The second order upper bound for the ground energy of a Bose gas. J. Stat. Phys. **136**, 453–503 (2009)
35. J. Yin, *Free Energies of Dilute Bose Gases*. Upper Bound. J. Stat. Phys. **141**, 683–726 (2010)

Chapter 3
Quantum Coulomb Gases

Jan Philip Solovej

3.1 Introduction

Ordinary matter is composed of electrons (negatively charged) and nuclei (positively charged) interacting via electromagnetic forces. The electric potential between two such particles of charges Q and Q' located at r and r' in $\mathbb{R}^3$ is $QQ'/|r-r'|$. The Coulomb potential poses two difficulties: (1) The local singularity and (2) its long range. One has to understand why the local singularity does not cause instabilities and why the long range does not have strong macroscopic effects. One of the first major triumphs of the theory of quantum mechanics is the explanation it gives of the stability of the hydrogen atom (and the complete description of its spectrum) and of other microscopic quantum Coulomb systems. It, surprisingly, took nearly 40 years before the question of stability of everyday macroscopic objects was even raised. The rigorous answer to the question came shortly thereafter in what came to be known as the *Theorem on Stability of Matter* proved first by Dyson and Lenard and later, in a more simple and transparent way by Lieb and Thirring. Since these seminal works, the proof of stability of matter has been extended to several different settings, including relativistic systems and systems in the presence of dynamic electromagnetic fields.

We will, in particular, discuss the importance of particle statistics, i.e., whether the particles are bosons or fermions. Using Bogolubov's theory for Bose gases Dyson concluded that charged bosonic particles would be macroscopically unstable. That Bogolubov's theory gives the correct description of charged Bose systems was proved in a series of papers by the author in collaboration with E.H. Lieb.

Having proved stability of matter the next question is whether one can establish the thermodynamics of charged systems, i.e., the existence of the thermodynamic

J.P. Solovej (✉)
Department of Mathematics, University of Copenhagen Universitetsparken 5,
2100 Copenhagen, Denmark
e-mail: solovej@math.ku.dk

V. Rivasseau et al., *Quantum Many Body Systems*, Lecture Notes in Mathematics 2051,
DOI 10.1007/978-3-642-29511-9_3, © Springer-Verlag Berlin Heidelberg 2012

limit. This was originally achieved by Lieb and Lebowitz. We will describe a new approach to the existence of the thermodynamic limit, which applies to many different quantum Coulomb systems, possibly in the presence of an underlying lattice structure. This generalizes earlier work of Fefferman.

An important theme in these notes is the use of functional inequalities, among which a prominent role is played by the Lieb–Thirring inequality.

Another important theme will be how to control the screening of the Coulomb potential.

The notes are organized as follows. In Sect. 3.2 we derive the classical Hamiltonian for charged particles interacting with electromagnetic fields. In Sect. 3.3 we discuss different ways of quantizing the system. We may quantize the particles and leave the fields unquantized or quantize both the particles and the fields. Moreover, we may consider the particles relativistically or classically. This will give a variety of different models. In Sect. 3.4 we discuss stability both in the sense of stability of individual atoms and in the sense of stability of macroscopic matter. Finally, in Sect. 3.5 we discuss situations where stability fails, in particular the case of bosonic matter.

3.2 Classical "Point" Charges

We consider N classical particles with charges $Q_1, \ldots, Q_N \in \mathbb{R}$ situated at points $r_1, \ldots, r_N \in \mathbb{R}^3$. Dimension 3 is of course the physical space dimension, but a natural questions would be whether the discussion of charged system could be generalized to other dimensions. It is however not entirely clear what the correct physical questions would be and we will there restrict the discussion to dimension 3.

We would ultimately like to consider the particles as point charges. Unfortunately this will however lead to divergencies. In order to avoid this we will initially assume that each particle is given by a charge distribution $Q_i \chi_R(r - r_i)$ where $\chi_R(r) = R^{-3}\chi(r/R)$ and $\chi \in C_0(\mathbb{R}^3)$ (a continuous compactly supported function) with $\int_{\mathbb{R}^3} \chi = 1$.

We will start the discussion of this system of charged particles from the Lagrangian of the particles and the electromagnetic field. It is[1] (using Gaussian units)

$$L_R(r_j, \mathbf{A}, V) = \sum_{j=1}^{N} \Big(T_i^*(\dot{r}_j) - Q_j \dot{r}_j \mathbf{A} * \chi_R(r_j) - Q_j V * \chi_R(r_j) \Big) + \frac{1}{8\pi} \int (|\partial_t \mathbf{A} + \nabla V|^2 - |\nabla \times \mathbf{A}|^2).$$

[1]Strictly speaking, even if we use a relativistic kinetic energy, this Lagrangian is not relativistically invariant. The reason is that we consider the particles as rigid bodies, which do not Lorenz contract as they move. We will here ignore this additional complication. The Lagrangian in the form given here is that of the Abraham model of charged particles [33].

where

- $\dot{r}_j$ denotes the velocity of particle j.
- $T_j^*(v)$ is the Legendre transform of the kinetic energy $T_j(p)$ as a function of momentum p. We assume that the kinetic energy functions T_j are convex functions on $\mathbb{R}^3$. For a non-relativistic particle of mass m_j we have $T_j(p) = \frac{1}{2m_j}p^2$ and thus $T_j^*(v) = \frac{1}{2}m_j v^2$ and for a relativistic particle it is $T_j(p) = \sqrt{p^2 + m_j^2} - m_j$ and hence $T_j^*(v) = m_j - m_j\sqrt{1-v^2}$. We are using units in which the speed of light is 1 (a convention we will use throughout these notes).
- $\mathbf{A}$ is the vector potential and the magnetic field is $\mathbf{B} = \nabla \times \mathbf{A}$.
- V is the electric potential. The electric field is

$$\mathbf{E} = -\partial_t \mathbf{A} - \nabla V.$$

In order to go to a Hamiltonian description we will choose Coulomb gauge or more precisely require that

$$\nabla \cdot \partial_t \mathbf{A} = 0,$$

i.e., we assume that $\nabla \cdot \mathbf{A}$ is independent of time. We will also for simplicity assume that $\mathbf{A}$ decays sufficiently fast that we are allowed to ignore boundary terms when integrating by parts.

With the Coulomb gauge condition we see that

$$\mathbf{E}_\perp = -\partial_t \mathbf{A}$$

is the divergence free part of the electric field. The total electric field is $\mathbf{E} = \mathbf{E}_\perp - \nabla V$ and we have

$$\int |\mathbf{E}|^2 = \int |\mathbf{E}_\perp|^2 + \int |\nabla V|^2.$$

The electric potential V is not a dynamic variable in the sense that that $\partial_t V = \dot{V}$ does not occur in L_R. The equation for V is

$$0 = \frac{\delta L_R}{\delta V} = -\frac{1}{4\pi}\Delta V - \sum_{j=1}^{N} Q_j \chi_R(r - r_j),$$

where we have used the Coulomb gauge condition. We recognize the equation for V as Gauss' law

$$4\pi \sum_{j=1}^{N} Q_j \chi_R(r - r_j) = -\Delta V = -\nabla \cdot (\partial_t \mathbf{A} + \nabla V) = \nabla \cdot \mathbf{E}.$$

In the Hamiltonian formalism this is a constraint equation. The solution is

$$V(r) = \sum_{j=1}^{N} Q_j \int \chi_R(r) * |r - r_j|^{-1} dr.$$

The canonical variable dual to r_j is

$$p_j = \nabla_{\dot{r}_j} L_R = \nabla T_j^*(\dot{r}_j) - Q_j \mathbf{A} * \chi_R(r_j).$$

The canonical variable dual to $\mathbf{A}$ is

$$\frac{\delta L_R}{\delta \partial_t \mathbf{A}} = -\frac{1}{4\pi}\partial_t \mathbf{A} = -(4\pi)^{-1}\mathbf{E}_\perp,$$

due to the Coulomb gauge condition. We then find the Hamiltonian function (where we assume that T_j is convex such that the double Legendre transform of $T_j^{**} = T_j$)

$$\begin{aligned}
H_R(r_j, p_j, \mathbf{A}, \mathbf{E}_\perp) &= \sum_{j=1}^{N} p_j \dot{r}_j - \frac{1}{4\pi}\int \partial_t \mathbf{A} \cdot \mathbf{E}_\perp - L_R(r_j, \mathbf{A}, V) \\
&= \sum_{j=1}^{N} T_j\big(p_j + Q_j \mathbf{A} * \chi_R(r_j)\big) + \sum_{j=1}^{N} Q_j V * \chi_R(r_j) \\
&\quad + \frac{1}{8\pi}\int (|\mathbf{E}_\perp|^2 + |\nabla \times \mathbf{A}|^2) - \frac{1}{8\pi}\int |\nabla V|^2 \\
&= \sum_{j=1}^{N} T_j\big(p_j + Q_j \mathbf{A} * \chi_R(r_j)\big) + \frac{1}{2}\sum_{j=1}^{N} Q_j V * \chi_R(r_j) \\
&\quad + \frac{1}{8\pi}\int (|\mathbf{E}_\perp|^2 + |\nabla \times \mathbf{A}|^2) \\
&= \sum_{j=1}^{N} T_j\big(p_j + Q_j \mathbf{A} * \chi_R(r_j)\big) \\
&\quad + \frac{1}{2}\sum_{j=1}^{N}\sum_{i=1}^{N} Q_j Q_i \iint \frac{\chi_R(r - r_i)\chi_R(r' - r_j)}{|r - r'|} dr dr' \\
&\quad + \frac{1}{8\pi}\int (|\mathbf{E}_\perp|^2 + |\mathbf{B}|^2).
\end{aligned}$$

If we subtract the (divergent) self-energy

$$\sum_{j=1}^{N} \frac{Q_j^2}{2} \iint \frac{\chi_R(r)\chi_R(r')}{|r-r'|} dr dr' = R^{-1} \sum_{j=1}^{N} \frac{Q_j^2}{2} \iint \frac{\chi(r)\chi(r')}{|r-r'|} dr dr'$$

and assume that $\mathbf{A}$ is continuous at r_j, we see that as R tends to zero H_R converges pointwise to the Hamiltonian

$$H(r_j, p_j, \mathbf{A}, \mathbf{E}_\perp) = \sum_{j=1}^{N} T_j(p_j + Q_j \mathbf{A}(r_j)) + \sum_{1 \le i < j \le N} \frac{Q_i Q_j}{|r_i - r_j|} + \frac{1}{8\pi} \int (|\mathbf{E}_\perp|^2 + |\mathbf{B}|^2). \tag{3.1}$$

Here r_j and p_j are dual canonical variables as are the field variables $\mathbf{A}$ and $-(4\pi)^{-1}\mathbf{E}_\perp$. Unfortunately, this Hamiltonian is only formal and suffers from singularities (the field $\mathbf{A}$ that solves Hamilton's equations will be singular at r_j) leading to severe difficulties in describing the motion of classical *point* charges.

If external fields V_{ex} and $\mathbf{A}_{\text{ex}}$ are present, the energy is

$$\sum_{j=1}^{N} \Big(T_j(p_j + Q_j(\mathbf{A} + \mathbf{A}_{\text{ex}})(r_j)) + Q_j V_{\text{ex}}(r_j)\Big) + \sum_{1 \le i < j \le N} \frac{Q_i Q_j}{|r_i - r_j|} + \frac{1}{8\pi} \int (|\mathbf{E}_\perp|^2 + |\mathbf{B}|^2).$$

3.3 Charged Quantum Gases

We now discuss the quantization of the Hamiltonian of charged point particles. We emphasize that it has not been possible to define a fully relativistically invariant and causal theory of quantum electrodynamics (QED) and all the models we describe here are at best approximations to such a theory (if it exists). All models discussed here are mathematically well defined (except when otherwise stated explicitly).

There are several levels of quantization that may be considered.

- We can leave the fields $\mathbf{A}$ and $\mathbf{E}_\perp$ classical and quantize the particles, i.e., describe them by a square integrable wave function $\psi(r_1, \ldots, r_N)$.
- We may quantize the particles and the fields, i.e., describe the particles in terms of a wave function ψ and turn the fields into operator valued functions $\mathbf{A}$ and $\mathbf{E}_\perp$. This would require introducing some cut-off regularization in the fields.
- We may second quantize the particles, i.e., let also ψ be an operator valued function. This procedure is necessary if we consider relativistic particles described by the Dirac operator.

3.3.1 Quantized Particles and Classical Fields

The variables of the system are the 3-dimensional vector fields $\mathbf{A}, \mathbf{E}_\perp$ (assumed to satisfy appropriate regularity and decay properties, at least implying that $\mathbf{E}_\perp$ and $\mathbf{B} = \nabla \times \mathbf{A}$ are square integrable) and the wave function ψ, which is a normalized function in $\bigotimes_{j=1}^N [L^2(\Omega)]^{\nu_j}$, where $\Omega \subset \mathbb{R}^3$ (say an open set) and ν_j is a positive integer counting the number of internal degrees of freedom of particle j, (e.g. a particle of spin s would correspond to $\nu_j = 2s + 1$). We shall write

$$\psi = \psi(r_1, s_1, \ldots, r_N, s_N), \quad r_j \in \Omega,\ s_j = 1, \ldots, \nu_j.$$

The energy of the system is given by

$$\mathscr{E}_N(\psi, \mathbf{A}, \mathbf{E}) = \langle \psi, H_N(\mathbf{A}, \mathbf{E}_\perp)\psi \rangle, \tag{3.2}$$

where $\langle \psi, \phi \rangle$ refers to the inner product of $\psi, \phi \in \bigotimes_{j=1}^N [L^2(\Omega)]^{\nu_j}$ and H_N is the (unbounded) operator (depending on $\mathbf{A}$ and $\mathbf{E}_\perp$)

$$H_N(\mathbf{A}, \mathbf{E}_\perp) = \sum_{j=1}^N \Big(T_j\big(-i\nabla_j + Q_j(\mathbf{A} + \mathbf{A}_{\rm ex})(r_j)\big) + Q_j V_{\rm ex}(r_j) \Big)$$
$$+ \sum_{1 \le i < j \le N} \frac{Q_i Q_j}{|r_i - r_j|} + \frac{1}{8\pi} \int (|\mathbf{E}_\perp|^2 + |\mathbf{B}|^2). \tag{3.3}$$

We will throughout be using units in which the reduced Planck constant $\hbar = 1$. The last integral above acts as a ($\mathbf{A}$ and $\mathbf{E}_\perp$ dependent) scalar in the Hilbert space. The Hamiltonian $H(\mathbf{A}, \mathbf{E}_\perp)$ depends also on the exterior fields $V_{\rm ex}$ and $\mathbf{A}_{\rm ex}$, but we suppress this in the notation as these fields usually remain fixed. In fact, we will mostly, and unless otherwise explicitly stated, assume that the exterior fields vanish, i.e., $V_{\rm ex} = 0$ and $\mathbf{A}_{\rm ex} = 0$.

The expectation value $\mathscr{E}_N(\psi, \mathbf{A}, \mathbf{E}_\perp)$ is not defined for all ψ in the Hilbert space

$$\bigotimes_{j=1}^N [L^2(\Omega)]^{\nu_j}.$$

We will here avoid the discussion of domains of self-adjointness of the operator $H(\mathbf{A}, \mathbf{E}_\perp)$. We will instead restrict attention to ψ in the subspace of smooth functions with compact support, i.e, $[C_0^\infty(\Omega^N)]^{\nu_1 \cdots \nu_N} \subseteq \bigotimes_{j=1}^N [L^2(\Omega)]^{\nu_j}$.

One of the main issues we will discuss in these notes is the question of stability, i.e., whether $\mathscr{E}_N(\psi, \mathbf{A}, \mathbf{E}_\perp)$ is bounded from below independently of ψ (normalized), $\mathbf{A}$, and $\mathbf{E}_\perp$. If such a lower bound holds the operator $H(\mathbf{A}, \mathbf{E}_\perp)$ has a self-adjoint Friedrichs extension and we are actually making claims about this extension. From our point of view the only complication due to considering the restriction to C_0^∞ is that a possible ground state (a state achieving the lowest possible energy) is most likely not represented by an element in C_0^∞, but only by an element in the Friedrichs extended domain. We shall, however, not be concerned with the actual ground states, but only the energy, so we ignore this issue.

Since the three coordinates of $-i\nabla_j + Q_j\mathbf{A}(r_j)$ correspond to in general non-commuting operators, we must discuss the meaning of $T_j\big(-i\nabla_j + Q_j\mathbf{A}(r_j)\big)$. We will, in fact, only consider examples where the functions $T_j(p)$ can be written in terms of (possibly matrix-valued) polynomial expressions of p in such a way that the meaning of T_j (at least on a suitable domain) will be clear. The examples we will consider are:

- Non-relativistic kinetic energy operators, where $T_j(p) = (2m_j)^{-1}p^2$, i.e., the operator is

$$T_j\big(-i\nabla_j + Q_j\mathbf{A}(r_j)\big) = (2m_j)^{-1}(-i\nabla_j + Q_j\mathbf{A}(r_j))^2. \qquad (3.4)$$

  We will refer to particles with this kinetic energy as non-relativistic particles. This is the kinetic energy used when treating non-relativistic atoms and molecules or ordinary matter.
- Relativistic kinetic energy operators, where $T_j(p) = (p^2+m_j^2)^{1/2}-m_j$, i.e., the operator is

$$T_j\big(-i\nabla_j + Q_j\mathbf{A}(r_j)\big) = ((-i\nabla_j + Q_j\mathbf{A}(r_j))^2 + m_j^2)^{1/2} - m_j. \qquad (3.5)$$

  The square root of an operator is here defined in the spectral theoretic sense.[2] We will refer to particles with this kinetic energy as relativistic (or sometimes pseudo-relativistic) particles. Both relativistic and non-relativistic particles may have internal degrees of freedom corresponding to ν_j, being greater than one.
- The non-relativistic and relativistic Pauli-operators. These are operators acting on two-component vector valued functions given by inserting the operator

$$\boldsymbol{\sigma}_j \cdot (-i\nabla_j + Q_j\mathbf{A}(r_j))$$

  into the kinetic energy functions above. Here $\boldsymbol{\sigma} = (\sigma^1, \sigma^2, \sigma^3)$ is the vector of 2×2 Pauli matrices

$$\sigma^1 = \begin{pmatrix} 0 & 1 \\ 1 & 0 \end{pmatrix}, \quad \sigma^2 = \begin{pmatrix} 0 & -i \\ i & 0 \end{pmatrix}, \quad \sigma^3 = \begin{pmatrix} 1 & 0 \\ 0 & -1 \end{pmatrix}.$$

  (The subscript j on $\boldsymbol{\sigma}$ above indicates that it acts on the internal degrees of freedom of particle j.) Thus in this case $\nu_j = 2$. The resulting kinetic energy

[2] The operator inside the square root is defined as a self-adjoint operator by Friedrichs extending it from the domain of smooth functions with compact support.

operators are

$$T_j\big(-i\nabla_j + Q_j\mathbf{A}(r_j)\big) = (2m_j)^{-1}(\boldsymbol{\sigma}_j \cdot (-i\nabla_j + Q_j\mathbf{A}(r_j)))^2 \tag{3.6}$$

for non-relativistic Pauli particles and

$$T_j\big(-i\nabla_j + Q_j\mathbf{A}(r_j)\big) = \Big(\big(\boldsymbol{\sigma}_j \cdot (-i\nabla_j + Q_j\mathbf{A}(r_j))\big)^2 + m_j^2\Big)^{1/2} - m_j \tag{3.7}$$

for relativistic Pauli particles. For the Pauli operator we have the Lichnerowicz' formula

$$\big(\boldsymbol{\sigma}_j \cdot (-i\nabla_j + Q_j\mathbf{A}(r_j))\big)^2 = (-i\nabla_j + Q_j\mathbf{A}(r_j))^2 + Q_j\boldsymbol{\sigma}_j \cdot \mathbf{B}(r_j) \tag{3.8}$$

and we see that the Pauli operator includes the coupling of the particle spin to the magnetic field.

- We could also consider the 4×4 Dirac operator

$$T_j(p) = \boldsymbol{\alpha} \cdot p + m_j\beta,$$

i.e.,

$$T_j(-i\nabla_j + Q_j\mathbf{A}(r_j)) = \boldsymbol{\alpha}_j \cdot (-i\nabla_j + Q_j\mathbf{A}(r_j)) + m_j\beta_j$$

where $\boldsymbol{\alpha}$ and β are standard 4×4 Dirac matrices, e.g.,

$$\boldsymbol{\alpha} = \begin{pmatrix} 0 & \boldsymbol{\sigma} \\ \boldsymbol{\sigma} & 0 \end{pmatrix}, \quad \beta = \begin{pmatrix} I & 0 \\ 0 & -I \end{pmatrix}$$

using a 2×2-block notation. Thus in this case $\nu_j = 4$. In contrast to the other types of operators the Dirac operator, however, is not positive, in fact, not bounded below and we will therefore not be able to treat it unless we second quantize the particle fields. We will discuss this briefly below. A different approach to deal with the unboundedness from below of the Dirac operator is to restrict to the subspace of $L^2(\mathbb{R}^3)^4$ which corresponds to the positive spectral subspace of the Dirac operator. This approach is called the no-pair theory, but we will not discuss it further here.

3.3.2 *Statistics of Identical Particles*

Until now all particles have been considered as distinguishable, but if we have identical particles the issue of particle statistics plays an important role.

The N-particle space for N-identical particles moving in $\Omega \subset \mathbb{R}^3$ and with ν internal degrees of freedom is $\mathcal{H}_N = \bigotimes^N [L^2(\Omega)]^\nu$. On $\mathcal{H}_N$ we define the orthogonal projections $P_N^{\pm}$

$$(P_N^{\pm}\psi)(r_1, s_1, \ldots, r_N, s_N)$$
$$= \frac{1}{N!}\sum_{\sigma \in S_N}(\pm 1)^{\sigma}\psi(r_{\sigma^{-1}(1)}, s_{\sigma^{-1}(1)}, \ldots, r_{\sigma^{-1}(N)}, s_{\sigma^{-1}(N)}).$$

They project onto the symmetric (+) or antisymmetric (−) subspaces. We denote these subspaces $\mathscr{H}_N^{\pm} = P_N^{\pm}\mathscr{H}_N$. We will also use the notation $\mathscr{H}_N^{-} = \bigwedge^N [L^2(\Omega)]^{\nu}$.

In case of N identical particles, i.e., if the operators T_j and the charges $Q_j = Q$ are the same for all $j = 1, \ldots, N$, the Hamiltonian H_N in (3.3) maps the subspaces $\mathscr{H}_N^{\pm}$ to themselves and it makes sense to restrict to these subspaces. We will write

$$\mathscr{E}_N^{\pm}(\psi, \mathbf{A}, \mathbf{E}_{\perp}) \tag{3.9}$$

to emphasize that we restrict to $\psi \in \mathscr{H}_N^{\pm}$. In the symmetric case (+) we say that we have a system of N *bosons* in the antisymmetric case (−) we say that we have a system of N fermions.

It is of course also possible to have mixtures of several species of identical particles being fermions or bosons or even several species of identical particles together with a number of distinguishable particles. We leave it to the reader to work out the structure of the underlying Hilbert space and the Hamiltonian in the general case. We will look at specific examples later.

3.3.3 Grand Canonical Picture

It is often useful to consider a situation where the particle number is not specified at the outset, but where we would instead ask what the optimal particle number is in a given situation, e.g., what number of particles minimizes the energy. This picture is referred to as the *grand canonical picture*. The optimal particle number may if necessary be adjusted by adding a term μ times the particle number to the Hamiltonian. Such a parameter μ is called a *chemical potential*. We note that this is not the same as adding a constant to the exterior electric potential V_{ext} as such a constant will multiply the total charge of the system.

In order to treat variable particle number we define the bosonic or fermionic *Fock spaces*

$$\mathscr{F}^{\pm} = \mathscr{F}^{\pm}((L^2(\Omega))^{\nu}) = \bigoplus_{N=0}^{\infty} \mathscr{H}_N^{\pm},$$

with the convention that $\mathscr{H}_0^{\pm} = \mathbb{C}$. The element $1 \in \mathbb{C} = \mathscr{H}_0^{\pm}$ is referred to as the vacuum vector and will be denoted by $|\mathbf{0}\rangle$. For a normalized vector $\psi \in \mathscr{F}^{\pm}$ we may write $\psi = \bigoplus_{N=0}^{\infty}\psi_N$, where $\psi_N \in \mathscr{H}_N^{\pm}$ with $\sum_{N=0}^{\infty}\|\psi_N\|^2 = 1$. We say that such a vector represents a *grand canonical state* and we define the *grand canonical energy* (with chemical potential μ included)

$$\mathcal{E}^{\pm}(\mu, \psi, \mathbf{A}, \mathbf{E}_{\perp}) = \sum_{N=0}^{\infty} \mathcal{E}_N^{\pm}(\psi_N, \mathbf{A}, \mathbf{E}_{\perp}) + \mu N \|\psi_N\|^2. \tag{3.10}$$

As before this energy is not defined for all ψ, but we restrict to ψ corresponding to finitely many particles, i.e., $\psi = \bigoplus_{N=0}^{M} \psi_N$ for some finite integer M and where each ψ_N is in C_0^{∞}.

Again it is possible to consider several species of identical particles in the grand canonical picture. We leave it to the reader to write down the Hilbert space and the energy (see also below).

3.3.4 Second Quantization and Quantization of Fields

We shall here give a brief introduction to second quantization and discuss how to quantize particle fields and the electromagnetic fields.

For $f \in L^2(\Omega)^{\nu}$ we define the annihilation operator $\widetilde{a}(f) : \mathcal{H}_N \to \mathcal{H}_{N-1}$ for $N = 1, \ldots$ by

$$(\widetilde{a}(f)\psi)(r_1, s_1, \ldots, r_{N-1}, s_{N-1}) = \sqrt{N} \sum_{s_N=1}^{\nu} \int \overline{f(r_N, s_N)} \psi(r_1, s_1, \ldots, r_N, s_N) dr_N.$$

The adjoint of this operator is $\widetilde{a}^*(f) : \mathcal{H}_{N-1} \to \mathcal{H}_N$ given by

$$(\widetilde{a}^*(f)\psi)(r_1, s_n, \ldots, r_N, s_N) = \sqrt{N} f(r_N, s_N) \psi(r_1, s_1, \ldots, r_{N-1}, s_{N-1}).$$

We define the bosonic and fermionic annihilation operators $a_{\pm}(f) : \mathcal{H}_N^{\pm} \to \mathcal{H}_{N-1}^{\pm}$ as the restriction of $\widetilde{a}(f)$ to the respective subspaces, i.e., $a_{\pm}(f) = \widetilde{a}(f)_{|\mathcal{H}_N^{\pm}}$. The adjoints are $a_{\pm}^*(f) : \mathcal{H}_{N-1}^{\pm} \to \mathcal{H}_N^{\pm}$ given by $a_{\pm}^*(f) = P_N^{\pm} \widetilde{a}^*(f)_{|\mathcal{H}_{N-1}^{\pm}}$.

We may extend $a_{\pm}(f)$ and $a_{\pm}^*(f)$ to operators on the subspace of the Fock spaces $\mathcal{F}^{\pm}$ corresponding to finite particle numbers, i.e., $\mathrm{span} \cup_{M=0}^{\infty} \bigoplus_{N=0}^{M} \mathcal{H}_N^{\pm}$. They cannot be extended as bounded operators on the full Fock spaces.

The extended operators satisfy the famous commutation (+) and anti-commutation (−) relations

$$[a_{\pm}(f), a_{\pm}^*(g)]_{\pm} = \langle f, g \rangle_{L^2(\Omega)^{\nu}} I$$

where $[A, B]_{\pm} = AB \mp BA$ and I is the identity of Fock space (or rather its restriction to the subspace corresponding to finite particle numbers.

If $\{f_j\}$ is an orthonormal basis in $L^2(\Omega)^{\nu}$ we define the operator valued distributions

$$\phi_{\pm}(r, s) = \sum_{j=1}^{\infty} f_j(r, s) a_{\pm}(f_j), \quad \phi_{\pm}^*(r, s) = \sum_{j=1}^{\infty} \overline{f_j(r, s)} a_{\pm}^*(f_j),$$

allowing us to write

$$a_{\pm}(f)=\sum_{s=1}^{\nu}\int\overline{f(r,s)}\phi_{\pm}(r,s)dr,\quad a_{\pm}^{*}(f)=\sum_{s=1}^{\nu}\int f(r,s)\phi_{\pm}^{*}(r,s)dr.$$

for $f\in L^2(\Omega)^\nu$. Formally we have

$$[\phi_{\pm}(r,s),\phi_{\pm}^{*}(r',s')]_{\pm}=\delta_{ss'}\delta(r-r').$$

We refer to ϕ and ϕ^* as field operators.

Using these field operators we may write the grand canonical Hamiltonian $\bigoplus_{N=0}^{\infty}(H_N+\mu N)$, corresponding to identical fermions or bosons, formally as

$$\begin{aligned}&\bigoplus_{N=0}^{\infty}(H_N(\mathbf{A},\mathbf{E}_\perp)+\mu N)\\&=\sum_{s,s'=1}^{\nu}\int\phi_{\pm}^{*}(r,s)\left[T^{ss'}(-i\nabla_r+Q\mathbf{A}(r))+\mu\delta_{ss'}\right]\phi_{\pm}(r,s')dr\\&+\frac{1}{2}\sum_{ss'}\iint\phi_{\pm}^{*}(r,s)\phi_{\pm}^{*}(r',s')\frac{Q^2}{|r-r'|}\phi_{\pm}(r',s')\phi_{\pm}(r,s)drdr'\\&+\frac{1}{8\pi}\int(|\mathbf{E}_\perp|^2+|\mathbf{B}|^2),\end{aligned}$$

where $T^{ss'}$, $s,s'=1,\ldots,\nu$ refer to the matrix components of the kinetic energy operator. It is left as an exercise to the reader to check that the ordering of ϕ^* and ϕ exactly gives the correct Hamiltonian with no-self interactions.

In this formalism it is easy also to write down the grand canonical operator corresponding to K different species of either fermions or bosons. For $j=1,\ldots,K$ let T_j, Q_j, and ν_j represent the kinetic energy function, the charge, and the internal degrees of freedom of species j which is either a fermion or a boson. The relevant Hilbert space is $\mathscr{H}=\bigotimes_{j=1}^{K}\mathscr{F}_j([L^2(\Omega)]^{\nu_j})$ where $\mathscr{F}_j$ is the Fock space for species j. Denoting the field operators for species j by ϕ_j and ϕ_j^* the corresponding grand canonical Hamiltonian (with chemical potential μ included) is

$$\begin{aligned}&H(\mu,\mathbf{A},\mathbf{E}_\perp)\\&=\sum_{j=1}^{K}\sum_{s,s'=1}^{\nu_j}\int\phi_j^{*}(r,s)\left[T_j^{ss'}(-i\nabla_r+Q_j\mathbf{A}(r))+\mu\delta_{ss'}\right]\phi_j(r,s')dr\\&+\frac{1}{2}\sum_{i,j=1}^{K}\sum_{s=1}^{\nu_i}\sum_{s'=1}^{\nu_j}\iint\phi_i^{*}(r,s)\phi_j^{*}(r',s')\frac{Q_iQ_j}{|r-r'|}\phi_j(r',s')\phi_i(r,s)drdr'\\&+\frac{1}{8\pi}\int(|\mathbf{E}_\perp|^2+|\mathbf{B}|^2).\end{aligned}$$

The subspace on which $H(\mu, \mathbf{A}, \mathbf{E}_\perp)$ is defined is the space corresponding to finitely many particles and where the restriction to each particle sector is a smooth function with compact support. Although it is hopefully clear what this means it is rather complicated to write it down explicitly. For the convenience of the reader we will nevertheless do this now. The subspace of $\mathscr{H}$ corresponding to N_j particles of species j, , $j = 1, \ldots, K$ is $\bigotimes_{j=1}^K P_j \left(\bigotimes^{N_j} [L^2(\Omega)]^{\nu_j}\right)$, where P_j refers to the relevant projection corresponding to the statistics of species j. We may consider this space a subspace of $[L^2(\Omega^{N_1+\ldots+N_K})]^{\nu_1^{N_1} \cdots \nu_K^{N_k}}$. The subspace of smooth functions with compact support is $[C_0^\infty(\Omega^{N_1+\ldots+N_K})]^{\nu_1^{N_1} \cdots \nu_K^{N_k}}$. Thus the subspace on which we define $H(\mu, \mathbf{A}, \mathbf{E}_\perp)$ is

$$\mathscr{D} = \operatorname{span} \bigcup_{M_1,\ldots,M_j=0}^{\infty} \bigoplus_{N_1=0}^{M_1} \cdots \bigoplus_{N_K=0}^{M_K} \left(\bigotimes_{j=1}^{K} P_j \bigotimes^{N_j} L^2(\Omega)^{\nu_j} \right)$$
$$\times \bigcap [C_0^\infty(\Omega^{N_1+\ldots+N_K})]^{\nu_1^{N_1} \cdots \nu_K^{N_k}}. \tag{3.11}$$

The energy of the system with particles being in a state represented by $\psi \in \mathscr{D}$ is denoted

$$\mathscr{E}(\mu, \psi, \mathbf{A}, \mathbf{E}_\perp) = \langle \psi, H(\mu, \mathbf{A}, \mathbf{E}_\perp)\psi \rangle. \tag{3.12}$$

It is easy to check that this agrees with the definition (3.10) in the case of only one species.

3.3.5 *Quantization of the Electromagnetic Field*

We will briefly discuss how to quantize the electromagnetic field. We will remain in Coulomb gauge and quantize such that $\nabla \cdot \mathbf{A} = 0$. This is most conveniently done in momentum space.

For $k \in \mathbb{R}^3$ choose $e_1(k), e_2(k) \in R^3$ such that $e_1(k), e_2(k), k$ form an orthonormal basis. e_1, e_2 cannot be chosen continuously, but this will not cause problems for what we want to say.

Let $\phi(r, \lambda)$, $\lambda = 1, 2$ be a bosonic field operator with two internal degrees of freedom. They are field operators for the light quanta, i.e., photons. Define the Fourier transformed operators (Of course they are also simply bosonic field operators)

$$\widehat{\phi}(k, \lambda) = (2\pi)^{-3/2} \int e^{-ikr} \phi(r, \lambda) dr, \quad \widehat{\phi}^*(k, \lambda) = (2\pi)^{-3/2} \int e^{ikr} \phi^*(r, \lambda) dr.$$

We define the quantized magnetic vector potential as the operator valued distribution

$$\mathbf{A}(r) = (2\pi)^{-3/2} \sum_{\lambda=1,2} \int_{\mathbb{R}^3} \sqrt{\frac{2\pi}{|k|}} e_\lambda(k)(e^{ikr}\widehat{\phi}(k,\lambda) + e^{-ikr}\widehat{\phi}(k,\lambda))dk \quad (3.13)$$

and the transversal electric field as

$$\mathbf{E}_\perp(r) = i(2\pi)^{-1/2} \sum_{\lambda=1,2} \int_{\mathbb{R}^3} \sqrt{\frac{|k|}{2\pi}} e_\lambda(k)(e^{ikr}\widehat{\phi}(k,\lambda) - e^{-ikr}\widehat{\phi}(k,\lambda))dk. \quad (3.14)$$

We then find the commutator between the conjugate variables

$$\left[\mathbf{A}_i(r), -\frac{1}{4\pi}\mathbf{E}_{\perp,j}(r'r)\right]_+ = \mathbf{P}_{i,j}(r,r'),$$

where $\mathbf{P}(r,r')$ is the 3×3-matrix valued integral kernel of the projection in $L^2(\mathbb{R}^3)^3$ projecting onto divergence free vector fields.

A straightforward (formal) calculation gives for the field energy

$$\frac{1}{8\pi}\int_{\mathbb{R}^3} |\mathbf{E}_\perp(r)|^2 + |\nabla\times\mathbf{A}(r)|^2 dr = \frac{1}{2}\sum_\lambda \int_{\mathbb{R}^3} |k|(\widehat{\phi}^*(k,\lambda)\widehat{\phi}(k,\lambda)$$
$$+\widehat{\phi}(k,\lambda)\widehat{\phi}^*(k,\lambda))dk.$$

This expression however is infinite and we must normal order it to get a well-defined operator:

$$\sum_\lambda \int_{\mathbb{R}^3} |k|\widehat{\phi}^*(k,\lambda)\widehat{\phi}(k,\lambda)dk.$$

This is the field energy operator of the electromagnetic field on the Fock space $\mathscr{F}^+(L^2(\mathbb{R}^3)^2)$.

3.3.6 Non-relativistic QED

We may now write down the Hamiltonian of non-relativistic QED, i.e., of the quantized electromagnetic field coupled to quantized non-relativistic particles. The particles will be described by the non-relativistic kinetic energies (3.4) or (3.6), but since A is now an operator valued distribution, these operators will not make sense unless we again introduce the extended charge distribution of the particles. The grand canonical non-relativistic QED Hamiltonian for K species of identical particles is then (ignoring for simplicity the chemical potential)

$$
\begin{aligned}
H = & \sum_{j=1}^{K} \sum_{s,s'=1}^{\nu_j} \int \phi_j^*(r,s) T_j^{ss'}(-i\nabla_r + Q_j \mathbf{A} * \chi_R(r)) \phi_j(r,s') dr \\
& + \sum_{i,j=1}^{K} \frac{Q_i Q_j}{2} \sum_{s=1}^{\nu_i} \sum_{s'=1}^{\nu_j} \iint \phi_i^*(r,s) \phi_j^*(r',s') \frac{1}{|r-r'|} \phi_j(r',s') \phi_i(r,s) dr dr' \\
& + \sum_{\lambda} \int_{\mathbb{R}^3} |k| \widehat{\phi}^*(k,\lambda) \widehat{\phi}(k,\lambda) dk.
\end{aligned}
$$

This operator is defined on the Hilbert space

$$
\Big(\bigotimes_{j=1}^{K} \mathscr{F}_j([L^2(\Omega)]^{\nu_j}) \Big) \bigotimes \mathscr{F}^+([L^2(\mathbb{R}^3)]^2).
$$

The operators ϕ_j are field operators for the particles and ϕ is the field operator for the photons. The energy may be calculated in a state represented by a Ψ in the subspace of the Hilbert space consisting of C_0^∞ functions of finitely many particles and photons (we will not write this explicitly this time). The energy is denoted

$$
\mathscr{E}_{\mathrm{NRQED}}(\Psi) = \langle \Psi, H\Psi \rangle. \tag{3.15}
$$

As written now the model depends on the regularization parameter R. The limit as R tends to 0 is not well understood and will require at least to renormalize the bare mass and charges of the particles.

3.3.7 Relativistic QED Hamiltonian

As already emphasized a Hamiltonian (or for that matter any non-perturbative) formulation of QED is non-existent. Here we simply write down the formal expression for the Hamiltonian for the electron–positron field (with charge e) interacting with the electromagnetic field:

$$
\begin{aligned}
H_{\mathrm{QED}} = & \sum_{a,b=1}^{4} \int \phi_{\mathrm{e}}^*(r,a) (\boldsymbol{\alpha} \cdot (-i\nabla + e\mathbf{A}(r)) + m\beta)_{a,b} \phi_{\mathrm{e}}(r,b) dr \\
& + \frac{e^2}{8} \sum_{a,b=1}^{4} \iint \frac{\left[\phi_{\mathrm{e}}(a,r), \phi_{\mathrm{e}}^*(a,r)\right]_+ \left[\phi_{\mathrm{e}}(b,r'), \phi_{\mathrm{e}}^*(b,r')\right]_+}{|r-r'|} dr dr' \\
& + \sum_{\lambda} \int_{\mathbb{R}^3} |k| \widehat{\phi}^*(k,\lambda) \widehat{\phi}(k,\lambda) dk.
\end{aligned}
$$

Here ϕ_e refers to the fermionic field operator for the electron–positron field and ϕ is the bosonic field operator for the photon field. The operator $\mathbf{A}$ is given by (3.13). Note that we have not distinguished between electrons and positrons, but that the operator is written in a charge conjugation invariant way as the density is written as the commutator $\frac{1}{2}\sum_{a=1}^{4}\left[\phi_e(a,r),\phi_e^*(a,r)\right]_+$.

The operator H_{QED} is ill-defined unless regularizations are introduced and even in this case it is very difficult to analyze. The no-photon situation was studied in the mean-field approximation in [15].

3.4 Stability

In the previous section we discussed how to define the energy of states of charged quantum gases in different models.

We have introduced the fixed particle number (or canonical) energy $\mathscr{E}_N(\psi,\mathbf{A},\mathbf{E}_\perp)$ in (3.2) (or the bosonic or fermionic analogs in (3.9)) or the grand canonical energy $\mathscr{E}(\mu,\psi,\mathbf{A},\mathbf{E}_\perp)$ in (3.12). We also defined the non-relativistic QED energy $\mathscr{E}_{\mathrm{NRQED}}(\Psi)$ in (3.15).

We will say that a system is *stable of the first kind* or *canonically stable* if the energy $\mathscr{E}_N(\psi,\mathbf{A},\mathbf{E}_\perp)$ is bounded below independently of $\mathbf{A}$, $\mathbf{E}_\perp$, and normalized ψ. In this case we will call the infimum of $\mathscr{E}_N(\psi,\mathbf{A},\mathbf{E})$ the *ground state* energy regardless of whether an actual minimizer (a ground state) exists or not. Thus the canonical ground state energy of the system is

$$E_N(\Omega) = \inf\left\{\mathscr{E}_N(\psi,\mathbf{A},\mathbf{E})\,\middle|\,\psi\in\left(\bigotimes_{j=1}^{N}L^2(\Omega)^{\nu_j}\right)\cap C_0^\infty(\Omega^N)^{\nu_1\ldots\nu_N},\ \|\psi\|=1,\right.$$
$$\left.\mathbf{A},\mathbf{E}_\perp\in C_0^\infty(\mathbb{R}^3;\mathbb{R}^3)\right\}.$$

Note that we are restricting the particles to be in the set Ω whereas $\mathbf{A}$ and $\mathbf{E}$ are unrestricted vector fields in $\mathbb{R}^3$. It is immediate to see that we might take $\mathbf{E}_\perp = 0$ in the infimum, this will however not be the case for quantized fields below.

The ground state energy of course depends on the types of particles in the system. We are suppressing this dependence in order not to overburden the notation.

The ground state is the state of the system at absolute zero temperature. It is of course also of interest to study quantum gases at positive temperature corresponding to minimizing the *free energy* we shall however not do this here.

We could also have chosen to consider the purely static Coulomb potential and set $\mathbf{A} = 0$, but as we shall see the inclusion of $\mathbf{A}$ does not really change the treatment in the non-relativistic (and non-Pauli) case from the points of view discussed here.

We say that a system satisfies *stability of the second kind* or *stability of matter* if $N^{-1}E_N(\Omega)$ is bounded below independently of N for all (open or in some cases sufficiently regular) $\Omega \subset \mathbb{R}^3$. This is the version of stability mainly studied in [23].

We will here use a slightly stronger notion which we refer to as *grand canonical stability*. We define the *grand canonical ground state energy* as

$$E(\mu,\Omega)=\inf\Big\{\mathscr{E}(\mu,\psi,\mathbf{A},\mathbf{E}_\perp)\;\Big|\;\psi\in\mathscr{D},\ \|\psi\|=1,\ \mathbf{A},\mathbf{E}_\perp\in C_0^\infty(\mathbb{R}^3;\mathbb{R}^3)\Big\},$$

where $\mathscr{E}(\mu,\psi,\mathbf{A},\mathbf{E}_\perp)$ was defined in (3.12). It of course depends on the species of particles.

We say that a system is grand canonically stable (with chemical potential μ) if

$$\inf_{\Omega\subseteq\mathbb{R}^3}|\Omega|^{-1}E(\mu,\Omega)>-\infty.$$

The infimum here is over all open sets Ω with bounded volume $|\Omega|$ (or possibly sufficiently regular sets if necessary, but we will not consider such cases here).

The original proof of stability of matter is due to Dyson and Lenard [7, 8] and later by a simpler method by Lieb and Thirring [25]. We will present a proof of grand canonical stability in a simple case relying on a combination of the two approaches.

For grand canonically stable systems it is of interest to consider whether the *thermodynamic limit*

$$\lim_{\Omega\to\mathbb{R}^3}|\Omega|^{-1}E(\mu,\Omega) \tag{3.16}$$

exists. The limit $\Omega\to\mathbb{R}^3$ can be given a precise meaning in different ways. Here we shall simply take the simple situation of the family of scaled copies $L\Omega$ of a fixed set Ω and let the real parameter L tend to infinity.

3.4.1 Stability of the First Kind for Non-relativistic Particles

We shall here prove the stability of the first kind for non-relativistic particles, i.e., particles with the kinetic energy (3.4).

Theorem 3.1 (Non-relativistic stability of the first kind). *For all $\psi\in C_0^\infty(\Omega^N)^{\nu_1\cdots\nu_N}$ and all vector fields $\mathbf{A},\boldsymbol{E}\in C_0^\infty(\mathbb{R}^3;\mathbb{R}^3)$ we have*

$$\Big\langle\psi,\Big(\sum_{j=1}^N\frac{1}{2m_j}(-i\nabla_j+Q_j\mathbf{A}(r_j))^2+\sum_{1\le i<j\le N}\frac{Q_iQ_j}{|r_i-r_j|}$$
$$+\frac{1}{8\pi}\int(|\boldsymbol{E}_\perp|^2+|\boldsymbol{B}|^2)\Big)\psi\Big\rangle\ge -C\|\psi\|^2,$$

where the constant $C>0$ depends only on the number of particles N and their properties, i.e., on $\nu_1,\ldots,\nu_N\in\mathbb{N}$, $m_1,\ldots,m_N>0$ and $Q_1,\ldots,Q_N\in\mathbb{R}$.

This theorem follows easily from the diamagnetic inequality and the Sobolev inequality (see [21]).

Theorem 3.2 (Diamagnetic Sobolev inequality). *For all* $f \in C_0^\infty(\mathbb{R}^3)$ *and all* $\mathbf{A} \in C_0^\infty(\mathbb{R}^3;\mathbb{R}^3)$ *there is a constant* $C > 0$ *such that*

$$\int_{\mathbb{R}^3} |(-i\nabla + \mathbf{A})f|^2 \geq \int_{\mathbb{R}^3} |\nabla|f||^2 \geq C\Big(\int_{\mathbb{R}^3} |f|^6\Big)^{1/3}.$$

An immediate corollary of this result (using simply Hölder's inequality) is the following bound on one-body Schrödinger operators.

Corollary 3.1 (Lower bound on Schrödinger operator). *For all* $f \in C_0^\infty(\mathbb{R}^3)$, $\mathbf{A} \in C_0^\infty(\mathbb{R}^3;\mathbb{R}^3)$, $0 \leq V_1 \in L^{5/2}(\mathbb{R}^3)$, *and* $0 \leq V_2 \in L^\infty(\mathbb{R}^3)$ *we have*

$$\langle f, ((-i\nabla + \mathbf{A})^2 - V_1 - V_2)f\rangle_{L^2} \geq -C\Big(\int V_1^{5/2} + \|V_2\|_\infty\Big)\|f\|_{L^2}^2.$$

We leave it to the reader to prove Theorem 3.1 from this corollary and the observation that $|r|^{-1} \in L^{5/2}(\mathbb{R}^3) + L^\infty(\mathbb{R}^3)$.

The stability of the first kind holds even if the field energy

$$\frac{1}{8\pi}\int |\mathbf{E}_\perp|^2 + |\mathbf{B}|^2$$

is ignored. Moreover, it also holds if $\mathbf{A}$ is quantized, i.e., if we replace $\mathbf{A}(r)$ by the operator (3.13). This last statement follows since $\mathbf{A}(r)$ is a commuting family (indexed by r) and thus may be considered as a classical field.

3.4.2 Grand Canonical Stability

We turn to the question of grand canonical stability. We will study this in the simple special case of two species of identical fermions with opposite charges. For grand canonical stability it is not necessary that all particles are fermions. It is, in fact, enough that all particles with one sign of the charge, i.e., say, all negatively charged particles form a collection of finitely many species of fermions. Stability of matter in this more general setting was proved in [7, 8, 25] (see also [18]) and the case of grand canonical stability was treated in [17].

One of the main ingredients in the proof of grand canonical stability is the use of the celebrated Lieb–Thirring inequality [25] (see also [23]) which replaces the Sobolev inequality which we used in the proof of stability of the first kind.

Theorem 3.3 (Lieb–Thirring inequality). *Assume* $0 \leq V \in L^{5/2}(\mathbb{R}^3)$ *and* $\mathbf{A} \in C_0(\mathbb{R}^3;\mathbb{R}^3)$ *then for all* N *we have on the antisymmetric subspace* $\mathscr{H}_N^- = \bigwedge^N [L^2(\Omega)]^\nu$ *the operator inequality*

$$\sum_{j=1}^{N}\left(\frac{1}{2m}(-i\nabla_j + \mathbf{A}(r_j))^2 - V(r_j)\right) \geq -Cm^{3/2}\nu\int V^{5/2},$$

for a universal constant $C > 0$. *In particular, it is independent of the number* N *of particles.*

Note the apparent similarity between the Lieb–Thirring inequality and the Corollary 3.1, to the Sobolev inequality. The important difference is that Corollary 3.1 would only imply that

$$\sum_{j=1}^{N}\left(\frac{1}{2m}(-i\nabla_j + \mathbf{A}(r_j))^2 - V(r_j)\right) \geq -Cm^{3/2}N\int V^{5/2},$$

which, in fact, holds on all of $\mathcal{H}_N$ (left as an exercise for the reader). The lower bound with a constant independent of N holds only on the fermionic subspace.

The Lieb–Thirring inequality relates the energy of a gas of independent particles to the corresponding classical energy. The classical energy (ignoring internal degrees of freedom) would indeed be

$$\iint\limits_{\frac{1}{2m}(p^2+\mathbf{A}(r))^2 - V(r)\leq 0} \frac{1}{2m}(p^2 + \mathbf{A}(r))^2 - V(r)drdp = -\frac{8\pi}{15}m^{3/2}\int V^{5/2}.$$

As a consequence of the Lieb–Thirring inequality we have the following lower bound on the kinetic energy of N fermions confined to move in a bounded volume.

Corollary 3.2. *If the open set* Ω *has finite volume* $|\Omega|$ *then in* $\bigwedge^N[L^2(\Omega)]^\nu$ *we have a universal constant* $C > 0$ *such that*

$$\sum_{j=1}^{N}\frac{1}{2m}(-i\nabla_j + \mathbf{A}(r_j))^2 \geq Cm^{-1}\nu^{-2/3}N^{5/3}|\Omega|^{-2/3}. \tag{3.17}$$

Proof. If we use the Lieb–Thirring inequality with a constant potential V we obtain

$$\sum_{j=1}^{N}\frac{1}{2m}(-i\nabla_j + \mathbf{A}(r_j))^2 \geq NV - Cm^{3/2}\nu V^{5/2}|\Omega|$$

which gives the estimate above after optimization in V. □

We now consider the situation with two species of identical non-relativistic fermions with masses $m_\pm > 0$ and charges $\pm Q_\pm$ where $Q_\pm > 0$. For simplicity we assume that there are no internal degrees of freedom, i.e., $\nu_\pm = 1$. In this case the Hamiltonian with particle numbers $N_\pm$ for the two species is

$$H_{N_+,N_-} = \sum_{j=1}^{N_+} \frac{1}{2m_+}(-i\nabla_j + Q_+\mathbf{A}(r_j))^2 + \sum_{j=N_++1}^{N_++N_-} \frac{1}{2m_-}(-i\nabla_j - Q_-\mathbf{A}(r_j))^2 + V_C$$
$$+\frac{1}{8\pi}\int (|E_\perp|^2 + |B|^2)$$

where the Coulomb energy is

$$V_C = -\sum_{j=1}^{N_+}\sum_{i=N_++1}^{N_++N_-} \frac{Q_+Q_-}{|r_i - r_j|} + \sum_{1\le i<j\le N_+} \frac{Q_+^2}{|r_i - r_j|} + \sum_{N_+<i<j\le N_++N_-} \frac{Q_-^2}{|r_i - r_j|}.$$

Note that we have numbered the positively charged particles $1, \ldots, N_+$ and the negatively charged particles $N_+ + 1, \ldots, N_+ + N_-$. The Hamiltonian acts on the subspace

$$\mathscr{D} = \left(\bigwedge^{N_+} L^2(\Omega) \otimes \bigwedge^{N_+} L^2(\Omega)\right) \cap C_0^\infty(\Omega^{N_++N_-}).$$

Theorem 3.4 (Simple case of grand canonical stability). *The grand canonical energy in the finite volume set* $\Omega \subseteq \mathbb{R}^3$

$$E(\mu,\Omega) = \inf\Big\{\langle\psi, H_{N_+,N_-}\psi\rangle + \mu(N_+N_-) \mid \psi \in \mathscr{D},\ \|\psi\| = 1,$$
$$\mathbf{E}_\perp, \mathbf{A} \in C_0^\infty(\mathbb{R}^3;\mathbb{R}^3)\Big\}$$

satisfies the stability bound

$$E(\mu,\Omega) \ge -C(\mu, m_\pm, Q_\pm)|\Omega|,$$

for a constant $C(\mu, m_\pm, Q_\pm) > 0$ *depending only on* $\mu, m_\pm, Q_\pm$.

Proof. We define the distance from particle j to the nearest particle of the opposite charge, i.e.,

$$\delta_j = \delta_j(r_1, \ldots, r_{N_++N_-})$$
$$= \begin{cases} \min_{i=N_++1,\ldots,N_++N_-} |r_i - r_j|, & \text{if } j = 1, \ldots, N_+ \\ \min_{i=1,\ldots,N_+} |r_i - r_j|, & \text{if } j = N_+ + 1, , \ldots, N_+ + N_- \end{cases}.$$

Let $\chi_j = \frac{6}{\pi\delta_j^3}\mathbf{1}_{B(r_j,\delta_j/2)}$, where $B(r_j, \delta_j/2)$ denotes the ball centered at r_j with radius $\delta_j/2$ and $\mathbf{1}_{B(r_j,\delta_j/2)}$ is its characteristic function. Note that $\int \chi_j = 1$.

We will use the following two observations:

Observation 1

$$\sum_{j=1}^{N_+}\sum_{i=N_++1}^{N_++N_-}\frac{Q_+Q_-}{|r_i-r_j|}=\sum_{j=1}^{N_+}\sum_{i=N_++1}^{N_++N_-}Q_+Q_-\iint\frac{\chi_j(r)\chi_i(r')}{|r-r'|}drdr'$$

Observation 2

$$\sum_{1\le i<j\le N_+}\frac{Q_+^2}{|r_i-r_j|}\ge\sum_{1\le i<j\le N_+}Q_+^2\iint\frac{\chi_j(r)\chi_i(r')}{|r-r'|}drdr'$$

and likewise for the Q_-^2-terms .

The observations follow from Newton's Theorem:

$$\frac{6}{\pi\delta^3}\int_{|r'|<\delta/2}|r-r'|^{-1}dr'=\begin{cases}|r|^{-1}, & \text{if } |r|>\delta/2\\ \delta^{-1}(3-4|r|^2\delta^{-2}), & \text{if } |r|<\delta/2\end{cases}\le|r|^{-1}.$$

From the two observations above we arrive at the following lower bound on the Coulomb energy

$$V_C\ge\frac{1}{2}\iint\frac{\rho(r)\rho(r')}{|r-r'|}drdr'-\frac{12}{5}\sum_{j=1}^{N_+}Q_+^2\delta_j^{-1}-\frac{12}{5}\sum_{j=N_++1}^{N_++N_-}Q_-^2\delta_j^{-1},$$

where we introduced the smeared charge density

$$\rho(r)=\sum_{j=1}^{N_+}Q_+\chi_j(r)-\sum_{j=N_++1}^{N_++N_-}Q_-\chi_j(r)$$

and used that

$$\iint\frac{\chi_j(r)\chi_j(r')}{|r-r'|}drdr'=\frac{12}{5}\delta_j^{-1}.$$

Using now the positive type (i.e., positivity of the Fourier transform) of the Coulomb kernel we find

$$V_C\ge-\frac{12}{5}\sum_{j=1}^{N_+}Q_+^2\delta_j^{-1}-\frac{12}{5}\sum_{j=N_++1}^{N_++N_-}Q_-^2\delta_j^{-1}.$$

A similar application of the positive type of the Coulomb kernel goes back to an early paper of Onsager [31], who might have been the first to address the issue of grand canonical stability. Better lower bounds on the Coulomb energy can be derived by more sophisticated use of the same ideas (see e.g. [1,23,26]).

We are led to the following lower bound on the Hamiltonian

$$H_{N_+,N_-} \geq H_{N_+} + H_{N_-} + \mu(N_+ + N_-)$$

where

$$H_{N_+} = \sum_{j=1}^{N_+} \frac{1}{2m_+}(-i\nabla_j + Q_+\mathbf{A}(r_j))^2 - \frac{12}{5}\sum_{j=1}^{N_+} Q_+^2 \delta_j^{-1}$$

and likewise for H_{N_-}. Observe now that for $j = 1, \ldots, N_+$ the length δ_j depends on the position r_j and the positions $r_{N_+ +1}, \ldots, r_{N_+ + N_-} R$ of the negatively charged particles but not on the positions of the other positively charged particles. In other words we may write

$$-\frac{12}{5}\sum_{j=1}^{N_+} Q_+^2 \delta_j^{-1} = -\frac{12}{5}\sum_{j=1}^{N_+} Q_+^2 \delta(r_j)^{-1},$$

where $\delta(r) = \min_{i=N_+ +1,\ldots,N_+ + N_-} |r_i - r|$. We thus have a potential parameterized by the positions of the negatively charged particles. This observation allows us to use the Lieb–Thirring inequality Theorem 3.3. If we choose a parameter R (to be optimized over) and divide the space into the region where $\delta(r) < R$ (a union of N_- possibly intersecting balls of radius R) and $\delta(r) > R$ we obtain from the Lieb–Thirring inequality

$$\begin{aligned} H_{N_+} &\geq \frac{1}{2}\sum_{j=1}^{N_+} \frac{1}{2m_+}(-i\nabla_j + Q_+\mathbf{A}(r_j))^2 \\ &\quad - C Q_+^5 m_+^{3/2}\Big(N_- \int_{|r|<R} |r|^{-5/2} dr + R^{-5/2}|\Omega|\Big) \\ &\geq C m_+^{-1} N_+^{5/3} |\Omega|^{-2/3} - C Q_+^5 m_+^{3/2}(N_- R^{1/2} + |\Omega| R^{-5/2}) \\ &= C m_+^{-1} N_+^{5/3} |\Omega|^{-2/3} - C Q_+^5 m_+^{3/2} N_-^{5/6} |\Omega|^{1/6}, \end{aligned}$$

where we saved half of the kinetic energy in the first inequality and estimated it by Corollary 3.2 in the second inequality. Finally, we optimized over the parameter $R > 0$. Since the corresponding estimate holds for H_{N_-} we finally get the lower bound

$$\begin{aligned} H_{N_+,N_-} &\geq C m_+^{-1} N_+^{5/3} |\Omega|^{-2/3} + C m_-^{-1} N_-^{5/3} |\Omega|^{-2/3} \\ &\quad - C Q_+^5 m_+^{3/2} N_-^{5/6} |\Omega|^{1/6} - C Q_-^5 m_-^{3/2} N_+^{5/6} |\Omega|^{1/6} + \mu(N_+ + N_-) \\ &\geq -C(\mu, m_\pm, Q_\pm)|\Omega|, \end{aligned}$$

where we have minimized in $N_\pm$. We leave it to the reader to determine the exact form of the constant $C(\mu, m_\pm, Q_\pm)$. □

The same proof would work also if periodic external electric and magnetic fields were present, e.g., a situation describing a crystal structure.

As should also be clear from the proof the field energy

$$\frac{1}{8\pi}\int |\mathbf{E}_\perp|^2 + |\mathbf{B}|^2$$

plays no role for stability in the present case. Moreover, as in the case discussed for stability of the first kind we could also have considered $\mathbf{A}$ quantized.

3.4.3 Existence of the Thermodynamic Limit

We will briefly discuss existence of the thermodynamic limit (3.16). This was first proved by Lieb and Lebowitz [20] for the case of several species of particles where all the species of, say, negatively charged particles are fermions. The method does not allow for an exterior periodic potential or magnetic field. In particular, the method does work in the case where the nuclei are confined to a periodic crystal arrangement. This case was later treated by Fefferman in [10]. In [16, 17] an abstract method was developed to conclude existence of thermodynamic limits for Coulomb systems in great generality including periodic background potentials.

Indeed, the method relies on establishing general abstract properties of the energy function that implies existence of the thermodynamic limit.

We will just give a brief overview of the method. For the details and more precise definitions and assumptions we refer to [16, 17].

Let $\mathscr{M} = \{\Omega \subset \mathbb{R}^3 \text{ open and bounded}\}$ and consider a map $E : \mathscr{M} \to \mathbb{R}$ with the following properties. Given a function $\alpha : [0, \infty)$ with $\lim_{\ell\to\infty}\alpha(\ell) = 0$, a subset $\mathscr{R} \subseteq \mathscr{M}$ of sufficiently regular sets, constants $\kappa, \delta > 0$, and a reference set $\triangle \in \mathscr{R}$, such that

(A1) *(Normalization).* $E(\emptyset) = 0$.
(A2) *(Stability).* $\forall \Omega \in \mathscr{M}$, $E(\Omega) \geq -\kappa|\Omega|$.
(A3) *(Translation Invariance).* $\forall \Omega \in \mathscr{R}$, $\forall z \in \mathbb{Z}^3$, $E(\Omega + z) = E(\Omega)$.
(A4) *(Continuity).* $\forall \Omega, \Omega' \in \mathscr{R}$, with $\Omega' \subseteq \Omega$ and $\mathrm{d}(\partial\Omega, \partial\Omega') > \delta$,
$$E(\Omega) \leq E(\Omega') + \kappa|\Omega \setminus \Omega'| + |\Omega|\alpha(|\Omega|).$$
(A5) *(Subaverage Property).* For all $\Omega \in \mathscr{M}$, we have

$$E(\Omega) \geq \frac{1}{|\ell\triangle|}\int_{\mathbb{R}^3 \rtimes SO(3)} E\big(\Omega \cap g\cdot(\ell\triangle)\big)\, d\lambda(g) - |\Omega|_{\mathrm{r}}\,\alpha(\ell) \tag{3.18}$$

where $d\lambda$ is the Haar-measure of $\mathbb{R}^3 \rtimes SO(3)$, (the group of isometries of $\mathbb{R}^3$) and $|\Omega|_{\mathrm{r}} := \inf\{|\tilde{\Omega}|,\quad \Omega \subseteq \tilde{\Omega},\quad \tilde{\Omega} \in \mathscr{R}\}$ is a regularized volume of Ω.

If $E : \mathscr{M} \to \mathbb{R}$ satisfies (A1–A5) then it is not very difficult to show that the thermodynamic limit

$$\lim_{\ell\to\infty} |\ell\Delta|^{-1} E(g\ell\Delta)$$

exists for all $g \in \mathbb{R}^3 \rtimes SO(3)$, i.e., it exists for all rotations or translations of the reference set Δ. Under slightly more restrictive assumptions which we will not repeat here the limit holds for a very large class of regular sets.

We see that (A2) is grand canonical stability. The difficult property to establish for Coulomb systems is (A5). For Δ being a simplex it is a consequence of the following result of Graf and Schenker [14] generalizing a somewhat similar estimate by Conlon et al. [5]:

Theorem 3.5 (Graf–Schenker inequality). *Let Δ be a simplex in $\mathbb{R}^3$. There exists a constant C such that for any $N \in \mathbb{N}$, $Q_1, \ldots, Q_N \in \mathbb{R}$, $r_1, \ldots, r_N \in \mathbb{R}^3$ and any $\ell > 0$,*

$$\sum_{1\leq i<j\leq N} \frac{Q_i Q_j}{|r_i - r_j|} \geq \frac{1}{|\ell\Delta|} \int_{\mathbb{R}^3\rtimes SO(3)} \sum_{1\leq i<j\leq N} \frac{Q_i Q_j \mathbf{1}_{g\ell\Delta}(r_i)\mathbf{1}_{g\ell\Delta}(r_j)}{|r_i - r_j|} d\lambda(g)$$
$$-\frac{C}{\ell}\sum_{j=1}^{N} Q_j^2.$$

This inequality follows by proving that the function

$$F(r, r') = \int_{\mathbb{R}^3\rtimes SO(3)} \mathbf{1}_{g\ell\Delta}(r_i)\mathbf{1}_{g\ell\Delta}(r_j) d\lambda(g),$$

is of the form $F(r, r') = g(|r - r'|)$ where g is such that $|r|^{-1}(1 - g(|r|)$ has positive Fourier transform. Recall that for a function f of positive type

$$\sum_{1\leq i<j\leq N} Q_i Q_j f(r_i - r_j) \geq -\sum_{j=1}^{N} Q_j^2 f(0).$$

3.5 Instability

3.5.1 Examples of Instability of the First Kind

As an example of a system that can show instability of the first kind we consider two relativistic particles with masses $m_1 = m_2 = 1$ and charges $Q_1 = -1$ and $Q_2 = Q > 0$. The kinetic energy is given by (3.5) and we simply set $\mathbf{A} = 0$. Thus the Hamiltonian is

$$H = \sqrt{-\Delta_1 + 1} - 1 + \sqrt{-\Delta_2 + 1} - 1 - \frac{Q}{|r_1 - r_2|}$$

acting on the smooth compactly supported functions in $L^2(\mathbb{R}^3) \otimes L^2(\mathbb{R}^3)$. Let $\psi \in C_0^\infty(\mathbb{R}^6)$ be normalized, i.e., its square integral is one. and define $\psi_\ell(r_1, r_2) = \ell^{-3}\psi(r_1/\ell, r_2/\ell)$ for $\ell > 0$. Note that ψ_ℓ is still normalized for all ℓ. Then

$$\langle \psi_\ell, H\psi_\ell \rangle = \ell^{-1}\left\langle \psi, \left(\sqrt{-\Delta_1 + \ell^2} - \ell + \sqrt{-\Delta_2 + \ell^2} - \ell - \frac{Q}{|r_1 - r_2|}\right)\psi \right\rangle.$$

Thus if we let ℓ tend to zero

$$\lim_{\ell \to 0} \ell \langle \psi_\ell, H\psi_\ell \rangle = \left\langle \psi, \left(\sqrt{-\Delta_1} + \sqrt{-\Delta_2} - \frac{Q}{|r_1 - r_2|}\right)\psi \right\rangle.$$

If Q is large enough we find that the right side is negative and hence for such a Q

$$\lim_{\ell \to 0} \langle \psi_\ell, H\psi_\ell \rangle = -\infty$$

and the system is not stable of the first kind.

On the other hand, if the negatively charged particles belong to a finite number of fermionic species and if the number of fermionic species, the maximal negative charge and the maximal positive charge satisfy appropriate bounds, then stability of matter holds [4, 12, 23, 26, 28].

A similar situation happens for non-relativistic particles interacting with magnetic fields according to the Pauli operator (3.6). Consider as an example a Hamiltonian for two particles of mass $m = 1$ and charges $Q_1 = Q > 0$ and $Q_2 = -Q < 0$:

$$H(\mathbf{A}) = \frac{1}{2}(\sigma \cdot (-i\nabla_1 - Q\mathbf{A}(r_1)))^2 + \frac{1}{2}(\sigma \cdot (-i\nabla_2 + Q\mathbf{A}(r_2)))^2 - \frac{Q^2}{|r_1 - r_2|}$$
$$+ \frac{1}{8\pi}\int |\nabla \otimes \mathbf{A}|^2,$$

where we have chosen $\mathbf{E} = 0$ (which is the energetically best choice). The instability in this case relies on the existence (see [9, 30]) of a non-zero $\widetilde{\psi} \in L^2(\mathbb{R}^3)$ and a magnetic field $\widetilde{\mathbf{A}}$ with $\int |\nabla \otimes \widetilde{\mathbf{A}}|^2 < \infty$ such that

$$\frac{1}{2}(\sigma \cdot (-i\nabla_1 - \widetilde{\mathbf{A}}(r_1)))^2 \widetilde{\psi} = 0.$$

We may assume that $\widetilde{\psi}$ is normalized. If for $\ell > 0$ we set

$$\psi_\ell(r_1, r_2) = \ell^{-3}\widetilde{\psi}(r_1/\ell)\widetilde{\psi}(r_2/\ell)$$

(which is also normalized) and $\mathbf{A}_\ell(r) = (Q\ell)^{-1}\widetilde{\mathbf{A}}(r/\ell)$ we obtain for the energy expectation

$$\ell\langle\psi_\ell, H(\mathbf{A}_\ell)\psi_\ell\rangle = -\left\langle\psi_{\ell=1}, \frac{Q^2}{|r_1 - r_2|}\psi_{\ell=1}\right\rangle + \frac{1}{8\pi Q^2}\int |\nabla\otimes\widetilde{\mathbf{A}}|^2.$$

Again we see that if Q is large enough the right side is negative and hence for such a Q we have as before $\lim_{\ell\to\infty}\langle\psi_\ell, H(\mathbf{A}_\ell)\psi_\ell\rangle = -\infty$. As for the relativistic case stability of matter also holds in this case under appropriate conditions [11,27]. This problem with a quantized field has been treated in [3, 13], the relativistic case with classical fields is considered in [29], and the relativistic case with quantized field in [22].

3.5.2 *Fermionic Instability of the Second Kind*

As the final topic of these notes we will discuss instability of the second kind.

We will first make a very simple general remark about instability of many-body systems with attractive interactions which has nothing to do with charged systems and holds even for fermions.

Theorem 3.6 (Fermionic instability for attractive 2-body potentials). *Assume that the potential $W : \mathbb{R}^n \to \mathbb{R}$ satisfies $W(r) \leq -c < 0$ for all r in a ball around the origin. Consider the N-body operator*

$$H_N = \sum_{j=1}^{N} -\frac{1}{2}\Delta_j + \sum_{1\leq i<j\leq N} W(r_i - r_j)$$

acting in the fermionic Hilbert space $\bigwedge^N L^2(\mathbb{R}^n)$. If $n \geq 3$ then H_N cannot be stable of the second kind, i.e., we can find a sequence of normalized vectors $\psi_N \in \bigwedge^N L^2(\mathbb{R}^n)$ such that

$$\lim_{N\to\infty} N^{-1}\langle\psi_N, H_N\psi_N\rangle = -\infty.$$

Proof. Assume that $W(r) \leq -c < 0$ on the ball of radius R centered at the origin. Define ψ_N as the (normalized) Slater determinant

$$\psi_N(r_1,\ldots,r_N) = (N!)^{-1/2}\det(u_i(r_j))_{i,j=1}^{N}$$

where u_j, $j = 1,\ldots,N$ are orthonormalized eigenfunctions corresponding to the N lowest eigenvalues of the negative Laplacian with Dirichlet boundary conditions for the largest cube centered at the origin and contained in the ball of radius R. We extend the functions to be 0 outside the cube. The functions u_j are explicit and can be written in terms of sines and cosines. It is a simple straightforward calculation to show that in all dimensions n there is a constant C_n such that

$$\langle \psi_N, \sum_{j=1}^{N} (-\frac{1}{2}\Delta_j)\psi_N \rangle \leq C_n N^{(n+2)/n} R^{-2}.$$

Comparing with Corollary 3.2 (written for the case $n = 3$) we see that there is always a similar lower bound.

Thus

$$N^{-1}\langle \psi_N, H_N \psi_N \rangle \leq C_n N^{2/n} R^{-2} - \frac{1}{2}(N-1)c.$$

We see that instability occurs when $n > 2$. □

3.5.3 *Instability of Bosonic Matter*

For matter consisting of charged particles we have discussed that the fermionic property ensures grand canonical stability. In this final section we will show that the fermionic property is indeed a necessity as stability fails for bosons.

We consider two species of bosons with masses $m_\pm = 1$, $Q_+ = -Q_- = 1$, $\mathbf{A} = \mathbf{E}_\perp = 0$. We describe them by the standard Schrödinger kinetic energy (3.4). If we have N_+ positively charged particles and N_- negatively charged particles we may write the Hamiltonian as

$$H_{N_+,N_-} = \sum_{j=1}^{N_+ + N_-} -\frac{1}{2}\Delta_j + \sum_{1\leq i<j\leq N_+ + N_-} \frac{e_i e_j}{|r_i - r_j|},$$

where $e_j = 1$ if $j = 1, \ldots, N_+$ and $e_j = -1$ if $j = N_+ + 1, \ldots, N_+ + N_-$. The Hilbert space is

$$\mathcal{H}_{N_+,N_-} = P^+_{N_+} \bigotimes^{N_+} L^2(\mathbb{R}^3) \otimes P^+_{N_-} \bigotimes^{N_-} L^2(\mathbb{R}^3).$$

This system is not stable of the second kind, in fact, the energy behaves like the number of particles to the $7/5$-th power. The following precise asymptotics was conjectured by Dyson in [6].

Theorem 3.7 (Dyson's formula). *Let*

$$E(N) = \inf_{N_+ + N_- = N} \inf\{\langle \psi, H_{N_+,N_-}\psi \rangle \,|\, \psi \in \mathcal{H}_{N_+,N_-} \cap C_0^\infty(\mathbb{R}^{3(N_+ + N_-)}), \|\psi\|=1\}$$

then as $N \to \infty$

$$\lim_{N\to\infty} \frac{E(N)}{N^{7/5}} = \inf\left\{ \tfrac{1}{2}\int_{\mathbb{R}^3} |\nabla\Phi|^2 - I_0 \int_{\mathbb{R}^3} \Phi^{5/2} \,\middle|\, 0 \le \Phi,\ \int_{\mathbb{R}^3} \Phi^2 = 1 \right\}, \quad (3.19)$$

with I_0 *given by*

$$I_0 = (2/\pi)^{3/4} \int_0^\infty 1 + x^4 - x^2\left(x^4+2\right)^{1/2} dx = \frac{4^{5/4}\Gamma(3/4)}{5\pi^{1/4}\Gamma(5/4)}. \quad (3.20)$$

From the Sobolev inequality (Theorem 3.2) it follows that the inf on the right of (3.19) is finite. In [6] Dyson proved an upper bound on $E(N)$ of the form $-cN^{7/5}$ and thus indeed proved the instability of the second kind. In [5] a lower bound of the form $-CN^{7/5}$ was established thus concluding that $7/5$ is the correct power. The theorem was finally proved in [24, 32]. In [19] Lieb proved that if the positively charged particles have infinite mass then the energy is much smaller, indeed, bounded above by $-CN^{5/3}$ a corresponding lower bound had already been proved in [7, 8].

The proof of Theorem 3.7 relies on an application of Bogolubov's theory of superfluidity [2]. The charged system, in fact, forms a superfluid state.

Dyson's formula (3.19) is proved by establishing the corresponding two inequalities. Establishing the lower bound is technically very involved and is beyond the scope of these notes. It is the content of the paper [24]. We will here give a brief sketch of the proof of the upper bound from [32]. The upper bound is proved by finding an appropriate trial state. Here we are guided by Bogolubov's theory.

It turns out that it is significantly easier to write down a grand canonical trial state than a canonical state. We are, however, interested in a canonical state. This will not be a serious problem as we will eventually be able to show that the state we construct is sharply peaked around the average particle number. We will ignore this point here and simply work with the grand canonical state. We refer the reader to [32] for details.

Another simplification is to consider the two species of bosons as one species with two internal degrees of freedom corresponding to the two signs of the charge. Constructing a trial state in this space will correspond to averaging over states with different numbers of positively and negatively charged particles.

We are thus considering the Fock space $\mathcal{F}^+ = \mathcal{F}^+(L^2(\mathbb{R}^3)^2)$. We write a function $f \in L^2(\mathbb{R}^3)^2$, as $f = f(r,e)$, where $e = \pm 1$ is the sign of the charge. Let $|\mathbf{0}\rangle$ be the vacuum vector in $\mathcal{F}^+$.

In constructing a bosonic trial state the first guess is to put all particles in the same one-particle state, i.e., to have a condensate. Let this state be represented by the (normalized) vector $\xi \in L^2(\mathbb{R}^3)^2$. Introduce first the normalized grand canonical vector

$$|\Xi\rangle = \exp\left(-\frac{N}{2} + \sqrt{N} a_+^*(\xi)\right)|\mathbf{0}\rangle = \sum_{n=0}^\infty e^{-N/2} \frac{N^{n/2}}{n!} a_+^*(\xi)^n |\mathbf{0}\rangle.$$

The corresponding state is an average over states with varying occupation in the condensate ξ. The average particle number in ξ is $\langle \Xi | a_+^*(\xi) a_+(\xi) | \Xi \rangle = N$ and the variance is also

$$\langle \Xi | (a_+^*(\xi) a_+(\xi))^2 | \Xi \rangle - \langle \Xi | a_+^*(\xi) a_+(\xi) | \Xi \rangle^2 = N.$$

Thus this state is peaked around particle number N with a standard deviation $\sqrt{N}$.

There is a unitary operator U on $\mathscr{F}^+$ such that

$$U^* a_+^*(f) U = a_+^*(f) + \sqrt{N} \langle \xi, f_\alpha \rangle.$$

Using this unitary we may also write $|\Xi\rangle = U|\mathbf{0}\rangle$.

A pure condensate like this will however not give the correct state. It is important to build pair excitations too. This is achieved as follows. Let $\{f_\alpha\}_{\alpha=0}^\infty$ be an orthonormal family in $L^2(\mathbb{R}^3)^2$ (they will represent the pair states). The normalized vector $\Psi \in \mathscr{F}^+$ representing our trial state may be abstractly written

$$\begin{aligned}\Psi &= \prod_{\alpha=0}^\infty (1-\lambda_\alpha^2)^{1/4} \exp\left(\sum_{\alpha=0}^\infty -\frac{\lambda_\alpha}{2} \left(a_+^*(f_\alpha) - \sqrt{N}\langle \xi, f_\alpha \rangle \right)^2 \right) |\Xi\rangle \qquad (3.21)\\ &= U \prod_{\alpha=0}^\infty (1-\lambda_\alpha^2)^{1/4} \exp\left(\sum_{\alpha=0}^\infty -\frac{\lambda_\alpha}{2} a_+^*(f_\alpha)^2 \right) |\mathbf{0}\rangle.\end{aligned}$$

We have introduced parameters $0 < \lambda_\alpha < 1$ with $\sum_{\alpha=0}^\infty \lambda_\alpha^2 < \infty$ to control the occupations in the pair states. For simplicity we will assume that ξ and $\{f_\alpha\}_{\alpha=0}^\infty$ are real functions.

We encode the information about the pair states in the positive semi-definite trace class operator on $L^2(\mathbb{R}^3)^2$

$$\gamma = \sum_{\alpha=0}^\infty \frac{\lambda_\alpha^2}{1-\lambda_\alpha^2} |f_\alpha\rangle\langle f_\alpha|. \qquad (3.22)$$

In terms of this operator a lengthy but straightforward calculation shows that

$$\left\langle \Psi, (a_+^*(f) - \sqrt{N}\langle \xi, f \rangle)(a_+(g) - \sqrt{N}\langle g, \xi \rangle)\Psi \right\rangle = \langle g, \gamma f \rangle,$$

(the inner product on the left is in $\mathscr{F}_+$ and the one on the right is in $L^2(\mathbb{R}^3)^2$) and $\left\langle \Psi, (a_+^*(f) - \sqrt{N}\langle \xi, f \rangle)\Psi \right\rangle = 0$. In particular,

$$\left\langle \Psi, a_+^*(f) a_+(g) \Psi \right\rangle = \langle g, (N|\xi\rangle\langle \xi| + \gamma) f \rangle.$$

Or equivalently using the field operators from Sect. 3.3.4

$$\big\langle \Psi, \phi(r,e)^*\phi(r',e')\Psi\big\rangle = N\xi(r,e)\xi(r',e') + \gamma(r,e;r',e') \tag{3.23}$$

where $\gamma(r,e;r',e')$ is the integral kernel of γ. Likewise,

$$\big\langle \Psi, a_+^*(f)a_+^*(g)\Psi\big\rangle = \Big\langle g, \Big(N|\xi\rangle\langle\xi| - \sqrt{\gamma(\gamma+1)}\Big)\, f\Big\rangle,$$

or

$$\big\langle \Psi, \phi(r,e)^*\phi(r',e')^*\Psi\big\rangle = N\xi(r,e)\xi(r',e') - \sqrt{\gamma(\gamma+1)}(r,e;r',e'). \tag{3.24}$$

Moreover, the state represented by Ψ satisfies Wick's formula, which for the 4-point function reads

$$\left\langle \Psi, \prod_{j=1}^{4}(a_+^\#(g_j) - \sqrt{N}\langle g_j,\xi\rangle^\#)\Psi\right\rangle$$
$$= \left\langle \Psi, \prod_{j=1,2}(a_+^\#(g_j) - \sqrt{N}\langle g_j,\xi\rangle^\#)\Psi\right\rangle\left\langle \Psi, \prod_{j=3,4}(a_+^\#(g_j) - \sqrt{N}\langle g_j,\xi\rangle^\#)\Psi\right\rangle$$
$$+\left\langle \Psi, \prod_{j=1,3}(a_+^\#(g_j) - \sqrt{N}\langle g_j,\xi\rangle^\#)\Psi\right\rangle\left\langle \Psi, \prod_{j=2,4}(a_+^\#(g_j) - \sqrt{N}\langle g_j,\xi\rangle^\#)\Psi\right\rangle$$
$$+\left\langle \Psi, \prod_{j=1,4}(a_+^\#(g_j) - \sqrt{N}\langle g_j,\xi\rangle^\#)\Psi\right\rangle\left\langle \Psi, \prod_{j=2,3}(a_+^\#(g_j) - \sqrt{N}\langle g_j,\xi\rangle^\#)\Psi\right\rangle.$$

Here # refers to either a $*$ (interpreted as complex conjugation on scalars) or no $*$. In particular, since ξ is real this gives

$$\begin{aligned}&\big\langle \Psi, (\phi(r,e)^* - \sqrt{N}\xi(r,e))(\phi(r',e')^* - \sqrt{N}\xi(r',e'))\\ &\quad\times(\phi(r',e') - \sqrt{N}\xi(r',e'))(\phi(r,e) - \sqrt{N}\xi(r,e))\Psi\big\rangle\\ &\quad= |\sqrt{\gamma(\gamma+1)}(r,e;r',e')|^2 + |\gamma(r,e;r',e')|^2\\ &\qquad+\gamma(r,e;r,e)\gamma(r',e';r',e')\end{aligned} \tag{3.25}$$

Armed with these identities we can calculate the expectation of the energy in the state represented by Ψ.

First we will explain, for the special case of the charged Bose system, how to choose the condensate function ξ and the trace class operator γ. More precisely, we will specify their charge dependence. We set

$$\xi(r,e) = \sqrt{\frac{1}{2}}\xi_0(r), \tag{3.26}$$

where ξ_0 is a real normalized function in $L^2(\mathbb{R}^3)$. Thus the condensate function does not depend on the charge. The operator γ on $L^2(\mathbb{R}^3)^2 = L^2(\mathbb{R}^3) \otimes \mathbb{C}^2$ will be chosen to have the form

$$\gamma = \gamma_0 \otimes \frac{1}{2}\begin{pmatrix} 1 & -1 \\ -1 & 1 \end{pmatrix},$$

where γ_0 is a positive trace-class operator on $L^2(\mathbb{R}^3)$. Put differently, the integral kernel of γ is chosen to be

$$\gamma(r,e;r',e') = \frac{1}{2}ee'\gamma_0(r;r'). \tag{3.27}$$

The charge part of this operator is a rank one operator and thus we also have

$$\sqrt{\gamma(\gamma+1)} = \sqrt{\gamma_0(\gamma_0+1)} \otimes \frac{1}{2}\begin{pmatrix} 1 & -1 \\ -1 & 1 \end{pmatrix}.$$

It is now straightforward to calculate the expectation of the Coulomb potential in the state represented by Ψ. From (3.23)–(3.27) we obtain

$$\begin{aligned}
&\left\langle \Psi, \bigoplus_{M=0}^{\infty} \sum_{1\leq i<j\leq M} \frac{e_i e_j}{|r_i - r_j|}\Psi \right\rangle \\
&= \left\langle \Psi, \frac{1}{2}\sum_{ee'=\pm 1} \iint ee'\phi(r,e)^*\phi(r',e')^*|r-r'|^{-1}\phi(r,e)\phi(r',e')\Psi \right\rangle \\
&= N\operatorname{Tr}_{L^2(\mathbb{R}^3)}\left(\mathscr{K}(\gamma_0 - \sqrt{\gamma_0(\gamma_0+1)})\right).
\end{aligned}$$

Here $\mathscr{K}$ is the operator with integral kernel

$$\mathscr{K}(r,r') = \xi_0(r)|r-r'|^{-1}\xi_0(r').$$

The total energy expectation is

$$\begin{aligned}
\left\langle \Psi, \bigoplus_{N_+,N_-=0}^{\infty} H_{N_+,N_-}\Psi \right\rangle &= \frac{N}{2}\int |\nabla\xi_0|^2 \\
&\quad + \frac{1}{2}\operatorname{Tr}(-\Delta\gamma_0) + N\operatorname{Tr}\Big(\mathscr{K}(\gamma_0 - \sqrt{\gamma_0(\gamma_0+1)})\Big).
\end{aligned}$$

The final step in the argument is to minimize the above expression over γ_0. More precisely this is done in a semiclassical approximation. We will only sketch this argument. The rigorous argument can again be found in [32]. We assume that γ_0 is the quantization of a classical symbol $f(r, p) \geq 0$. The semiclassical approximation to the energy is then

$$\frac{N}{2}\int |\nabla \xi_0|^2$$
$$+ (2\pi)^{-3} \iint \frac{p^2}{2} f(r,p) + 4\pi N |p|^{-2} \xi_0(r)^2 \big(f(r,p)$$
$$- \sqrt{f(r,p)(f(r,p)+1)}\big) dr dp.$$

Minimizing this expression over $f(r, p)$ and performing the p integration gives

$$\frac{N}{2}\int |\nabla \xi_0|^2 - I_0 N^{5/4} \int \xi_0(r)^{5/2} dr,$$

where I_0 is given in (3.20). If we introduce the rescaling $\xi_0(r) = N^{3/10}\Phi(N^{1/5}r)$, where Φ is also normalized then the energy expression above becomes

$$N^{7/5}\Big(\int |\nabla \Phi|^2 - I_0 \int \Phi^{5/2}\Big),$$

which is exactly the expression conjectured by Dyson for the energy.

Note that the instability is also reflected in the shrinking of the linear dimension of the state with increasing N. According to the scaling of ξ_0 above, the linear dimension of the state behaves like $\sim N^{-1/5}$.

Acknowledgement Many thanks to the organizers for the invitation to give these lectures and in particular to A. Giuliani for the financial support through the ERC starting grant CoMboS-239694.

References

1. R.J. Baxter, Inequalities for potentials of particle systems. Illinois J. Math. **24**(4), 645–652 (1980)
2. N. Bogolubov, On the theory of superfluidity. J. Phys. (USSR) **11**, 23 (1947)
3. L. Bugliaro, J. Fröhlich, G.M. Graf, Stability of quantum electrodynamics with nonrelativistic matter. Phys. Rev. Lett. **77**(17), 3494–3497 (1996)
4. J.G. Conlon, The ground state energy of a classical gas. Comm. Math. Phys. **94**(4), 439–458 (1984)
5. J.G. Conlon, E.H. Lieb, H.-T. Yau, The $N^{7/5}$ law for charged bosons. Comm. Math. Phys. **116**(3), 417–448 (1988)
6. F.J. Dyson, Ground state energy of a finite system of charged particles. J. Math. Phys. **8**, 1538–1545 (1967)

7. F.J. Dyson, A. Lenard, Stability of matter, I. J. Math. Phys. **8**, 423–434 (1967)
8. F.J. Dyson, A. Lenard, Stability of matter, II. J. Math. Phys. **9**, 698–711 (1968)
9. L. Erdős, J.P. Solovej, The kernel of Dirac operators on $\mathbf{S}^3$ and $\mathbf{R}^3$. Rev. Math. Phys. **13**(10), 1247–1280 (2001)
10. C. Fefferman, The thermodynamic limit for a crystal. Comm. Math. Phys. **98**, 289–311 (1985)
11. C. Fefferman, Stability of matter with magnetic fields. CRM Proc. Lect. Notes **12**, 119–133 (1997)
12. C. Fefferman, R. de la Llave, Relativistic stability of matter, I. Rev. Mat. Iberoamericana **2**, 119–213 (1986)
13. C. Fefferman, J. Fröhlich, G.M. Graf, Stability of nonrelativistic quantum mechanical matter coupled to the (ultraviolet cutoff) radiation field. Proc. Natl. Acad. Sci. USA **93**, 15009–15011 (1996); Stability of ultraviolet cutoff quantum electrodynamics with non-relativistic matter. Comm. Math. Phys. **190**, 309–330 (1997)
14. G.M. Graf, D. Schenker, On the molecular limit of Coulomb gases. Comm. Math. Phys. **174**(1), 215–227 (1995)
15. C. Hainzl, M. Lewin, J.P. Solovej, The mean-field approximation in quantum electrodynamics: The no-photon case. Comm. Pure Appl. Math. **60**, 546–596 (2007)
16. C. Hainzl, M. Lewin, J.P. Solovej, The thermodynamic limit of quantum Coulomb systems, Part I. General theory. Adv. Math. **221**, 454–487 (2009)
17. C. Hainzl, M. Lewin, J.P. Solovej, The thermodynamic limit of quantum Coulomb systems, Part II. Applications. Adv. Math. **221**, 488–546 (2009)
18. E.H. Lieb, The stability of matter. Rev. Mod. Phys. **48**, 553–569 (1976)
19. E.H. Lieb, The $N^{5/3}$ law for bosons. Phys. Lett. **70A**, 71–73 (1979)
20. E.H. Lieb, J.L. Lebowitz, The constitution of matter: existence of thermodynamics for systems composed of electrons and nuclei. Adv. Math. **9**, 316–398 (1972)
21. E.H. Lieb, M. Loss, in *Analysis*. Graduate Studies in Mathematics, vol. 14 (American Mathematical Society, Providence, 2001)
22. E.H. Lieb, M. Loss, Stability of a model of relativistic quantum electrodynamics. Comm. Math. Phys. **228**, 561–588 (2002)
23. E.H. Lieb, R. Seiringer, *The Stability of Matter in Quantum Mechanics* (Cambridge University Press, Cambridge, 2010)
24. E.H. Lieb, J.P. Solovej, Ground state energy of the two-component charged Bose gas. Comm. Math. Phys. **252**, 485–534 (2004)
25. E.H. Lieb, W.E. Thirring, Bound for the kinetic energy of fermions which proves the stability of matter. Phys. Rev. Lett. **35**, 687–689 (1975)
26. E.H. Lieb, H.-T. Yau, The stability and instability of relativistic matter. Comm. Math. Phys. **118**, 177–213 (1988)
27. E.H. Lieb, M. Loss, J.P. Solovej, Stability of matter in magnetic fields. Phys. Rev. Lett. **75**, 985–989 (1995)
28. E.H. Lieb, M. Loss, H. Siedentop, Stability of relativistic matter via Thomas-Fermi theory. Helv. Phys. Acta **69**, 974–984 (1996)
29. E.H. Lieb, H. Siedentop, J.P. Solovej, Stability and instability of relativistic electrons in classical electromagnetic fields. J. Stat. Phys. **89**, 37–59 (1997)
30. M. Loss, H.-T. Yau, Stability of Coulomb systems with magnetic fields, III. Zero energy bound states of the Pauli operator. Comm. Math. Phys. **104**, 283–290 (1986)
31. L. Onsager, Electrostatic interaction of molecules. J. Phys. Chem. **43**, 189–196 (1939)
32. J.P. Solovej, Upper bounds to the ground state energies of the one- and two-component charged Bose gases. Comm. Math. Phys. **266**(3), 797–818 (2006)
33. H. Spohn, *Dynamics of Charged Particles and Their Radiation Field* (Cambridge University Press, Cambridge, 2004)

Chapter 4
SUSY Statistical Mechanics and Random Band Matrices

Thomas Spencer

4.1 An Overview

The study of large random matrices in physics originated with the work of Eugene Wigner who used them to predict the energy level statistics of a large nucleus. He argued that because of the complex interactions taking place in the nucleus there should be a random matrix model with appropriate symmetries, whose eigenvalues would describe the energy level spacing statistics. Recent developments summarized in [19], give a rather complete description of the universality of eigenvalue spacings for the *mean field* Wigner matrices. Today, random matrices are studied in connection with many aspects of theoretical and mathematical physics. These include Anderson localization, number theory, generating functions for combinatorial enumeration, and low energy models of QCD. See [6] for an explanation of how random matrix theory is related to these topics and others.

The goal of these lectures is to present the basic ideas of *supersymmetric* (SUSY) statistical mechanics and its applications to the spectral theory of large Hermitian random matrices—especially *random band matrices*. There are many excellent reviews of SUSY and random matrices in the theoretical physics literature starting with the fundamental work of Efetov [17]. See for example [15,18,28,31,36,49,51]. This review will emphasize mathematical aspects of SUSY statistical mechanics and will try to explain ideas in their simplest form. We shall begin by first studying the average Green's function of $N \times N$ GUE matrices—Gaussian matrices whose distribution is invariant under unitary transformations. In this case the SUSY models can be expressed as integrals in just two real variables. The size of the matrix N appears only as a parameter. See Sects. 4.3–4.5.

The simple methods for GUE are then generalized to treat the more interesting case of *random band matrices*, RBM, for which much less is rigorously proved.

T. Spencer (✉)
School of Mathematics, Institute for Advanced Study, Einstein Drive, Princeton, NJ 08540, USA
e-mail: spencer@math.ias.edu

V. Rivasseau et al., *Quantum Many Body Systems*, Lecture Notes in Mathematics 2051, DOI 10.1007/978-3-642-29511-9_4, © Springer-Verlag Berlin Heidelberg 2012

Random band matrices H_{ij} are indexed by vertices i, j of a lattice $\mathbb{Z}^d$. Their matrix elements are random variables which are 0 or small for $|i - j| \geq W$ and hence these matrices reflect the geometry of the lattice. The parameter W will be referred to as the width of the band. As we vary W, random band matrices approximately interpolate between mean field $N \times N$ GUE or Wigner type matrix models where W = N and random Schrödinger matrices, $H = -\Delta + v_j$ on the lattice in which the randomness only appears in the potential v. Following Anderson, random Schrödinger matrices are used to model the dynamics of a quantum particle scattered by random impurities on a crystalline lattice. In Sect. 4.3 precise definitions of random band and random Schrödinger matrices are given and a qualitative relation between them is explained.

The key to learning about spectral properties of random matrices H lies in the study of averages of its Green's functions, $(E_\epsilon - H)^{-1}(j,k)$ where j,k belong to a lattice. We use the notation $A(j,k)$ to denote the matrix elements of the matrix A. The energy E typically lies inside the spectrum of H and $E_\epsilon = E - i\epsilon$ with $\epsilon > 0$. Efetov [17] showed that averages of products of Green's functions can be exactly expressed in terms of correlations of certain supersymmetric (SUSY) statistical mechanics ensembles. In SUSY statistical mechanics models, the spin or field at each lattice site has both real and Grassmann (anticommuting) components. See (4.20) below for the basic identity. These components appear symmetrically and thus the theory is called supersymmetric. Efetov's formalism builds on the foundational work of Wegner and Schäfer [41, 50] which used replicas instead of Grassmann variables. A brief review of Gaussian and Grassmann integration is given in Appendix 4.10. Although Grassmann variables are a key element of SUSY models, we shall frequently integrate them out so that the statistical mechanics involves only real integrals.

The first step in the SUSY approach to random band matrices is to define a SUSY statistical mechanics model whose correlations equal the averaged Green's functions. This is basically an algebraic step, but some analysis is needed to justify certain analytic continuations and changes of coordinates. This gives us a *dual representation* of the averaged Green's function in terms of SUSY statistical mechanics. In the case of Gaussian randomness, the average of the Green's function can be explicitly computed and produces a quartic action in the fields. The advantage of this dual representation is that many of the concepts of statistical mechanics such as phase transitions, saddle point analysis, cluster expansions and renormalization group methods [40], can then be used to analyze the behavior of Green's functions of random band matrices. Section 4.6 will review some results about phase transitions and symmetry breaking for classical statistical mechanics. This may help to give some perspective of how SUSY statistical mechanics is related to its classical counterpart.

The second step is to analyze correlations of the resulting SUSY statistical mechanics model. Typically these models have a formal non-compact symmetry. For large W, the functional integral is expected to be governed by a finite dimensional *saddle manifold*. This manifold is the orbit of a critical point under the symmetry group. The main mathematical challenge is to estimate the fluctuations about the manifold. Most (SUSY) lattice models are difficult to analyze rigorously due to the lack of a spectral gap and the absence of a positive integrand. These lectures will focus on some special cases for which the analysis can be rigorously carried out.

4.1.1 Green's Functions

Let H be a random matrix indexed by vertices $j \in \mathbb{Z}^d$. We can think of H as a random Schrödinger matrix or an infinite random band matrix of fixed width acting on $\ell_2(\mathbb{Z}^d)$. To obtain information about time evolution and eigenvectors of H we define the average of the square modulus of the Green's function

$$< |G(E_\epsilon : 0, j)|^2 > \equiv < |(E - i\epsilon - H)^{-1}(0, j)|^2 > \equiv \Gamma(E_\epsilon, j) \tag{4.1}$$

where $j \in \mathbb{Z}^d$, $\epsilon > 0$, and $< \cdot >$ denotes the average over the random variables of H. We wish to study $\Gamma(E_\epsilon, j)$ for very small $\epsilon > 0$. If for all small $\epsilon > 0$ we have

$$\Gamma(E_\epsilon, j) \leq C\epsilon^{-1} e^{-|j|/\ell} \tag{4.2}$$

then the eigenstates with eigenvalues near E are *localized*. This means the eigenstate ψ decays exponentially fast $|\psi_j|^2 \approx e^{-|j-c|/\ell}$, about some point c of the lattice with probability one. The length $\ell = \ell(E)$ is called the *localization length*.

Quantum diffusion near energy E corresponds to

$$\Gamma(E_\epsilon, j) \approx (-D(E)\Delta + \epsilon)^{-1}(0, j) \quad or \quad \hat{\Gamma}(E_\epsilon, p) \approx (D(E)p^2 + \epsilon)^{-1} \tag{4.3}$$

where $D(E)$ is the diffusion constant and Δ is the discrete lattice Laplacian and $|p|$ is small. Note that the right side is the Laplace transform of the heat kernel. When (4.3) holds with $D(E) \geq \delta > 0$, particles with energy near E are mobile and can conduct. The eigenstates ψ_j at these energies are *extended*. This means that if H is restricted to a large volume Λ, an $\ell_2(\Lambda)$ normalized eigenfunction ψ satisfies $|\psi_j|^2 \approx |\Lambda|^{-1}$ for all $j \in \Lambda$. A brief perturbative discussion of quantum diffusion is presented in Appendix 4.11. In the infinite volume limit, quantum diffusion implies absolutely continuous spectrum near E, whereas localization corresponds to dense point spectrum and the absence of conduction. Although the quantum diffusion in 3D is well studied in the physics literature, it remains a major open problem to establish it at a rigorous mathematical level.

The basic identity

$$Im < G(E_\epsilon; 0, 0) > = \epsilon \sum_j < |G(E_\epsilon; 0, j)|^2 > \tag{4.4}$$

is easily proved by applying the resolvent identity to $G - \bar{G}$. It reflects conservation of probability or unitarity and is some times referred to as a Ward identity. In

Sect. 4.5 we shall see that the left side of (4.4) is proportional to the density of states, $\rho(E)$. Thus the identity $\sum_j \Gamma(E_\epsilon, j) \propto \epsilon^{-1}\, \rho(E)$ always holds and explains the factor ϵ^{-1} in (4.2).

In the language of SUSY statistical mechanics, localization roughly corresponds to high temperature in which distant spins become exponentially decorrelated, whereas quantum diffusion corresponds to an ordered phase in which spins are aligned over long distances. For random band matrices, the inverse temperature is

$$\beta \approx W^2 \rho(E)^2 \tag{4.5}$$

where $\rho(E)$ is the density of states, and W is the band width.

For a more intuitive time dependent picture, let $U(j,t)$ be the solution to the Schrödinger equation on the lattice

$$i\partial_t\, U(j,t) = H\, U(j,t). \tag{4.6}$$

Suppose that at time 0, $U(j,0)$ is supported near the origin. For any unitary evolution we have conservation of probability: $\sum_j |U(j,t)|^2$ is constant in t. To measure the spread of U we define the average mean-square displacement by

$$R^2(t) = <\sum_j |U(j,t)|^2 |j|^2 > . \tag{4.7}$$

Quantum diffusion corresponds to $R^2 \approx Dt$, whereas if all states are uniformly localized then the wave packet U does not spread as time gets large, $R^2 \leq Const\, \ell^2$. The Green's function at energy E, essentially projects the initial wave function U onto a spectral subspace about E. Time is roughly proportional to $t \approx \epsilon^{-1}$.

Note that if $H = -\Delta$ is the lattice Laplacian on $\mathbb{Z}^d$, and there no disorder, the motion is *ballistic*: $R^2 \propto t^2$. The presence of disorder, i.e. randomness, should change the character of the motion through multiple scatterings. It is expected that in the presence of spatially uniform disorder, the motion is never more than diffusive, $R^2(t) \leq Ct^{2\alpha}$ with $0 \leq \alpha \leq 1/2$ with possible logarithmic corrections in the presence of spin-orbit coupling. However, in two or more dimensions, this is still a conjecture, even for $\alpha < 1$. It is easy to show that for the lattice random Schrödinger operator we have $R^2(t) \leq Ct^2$ for any potential.

4.1.2 Symmetry and the 1D SUSY Sigma Model

Symmetries of statistical mechanics models play a key role in the macroscopic behavior of correlations. In Sect. 4.6 we present a review of phase transitions, symmetry breaking and Goldstone modes (slow modes) in classical statistical mechanics. The SUSY lattice field action which arises in the study of (4.1) is invariant under a global hyperbolic $SU(1,1|2)$ symmetry. As mentioned earlier, the spin or field has both real and Grassmann components. The symmetry $SU(1,1|2)$ means that for the real components there exists a hyperbolic symmetry U(1, 1) preserving $|z_1|^2 - |z_2|^2$ as discussed in Sect. 4.7. On the other hand, the Grassmann variables (reviewed in Appendix 4.10) are invariant under a compact SU(2) symmetry. More details about these symmetries are in Sect. 4.9.

In both physics and mathematics, many statistical mechanics systems are studied in the *sigma model approximation*. In this approximation, spins take values in a symmetric space. The underlying symmetries and spatial structure of the original interaction are respected. The Ising model, the rotator, and the classical Heisenberg model are well known sigma models where the spin s_j lies in $\mathbb{R}^1$, $\mathbb{R}^2$, $\mathbb{R}^3$, respectively with the constraint $|s_j^2| = 1$. Thus they take values in the groups Z_2, S^1 and the symmetric space S^2. One can also consider the case where the spin is matrix valued. In the Efetov sigma models, the fields are 4×4 matrices with both real and Grassmann entries. It is expected that sigma models capture the qualitative physics of more complicated models with the same symmetry, see Sect. 4.6.

In a one dimensional chain of length L, the SUSY sigma model governing conductance has a simple expression first found in [16]. The Grassmann variables can be traced out and the resulting model is a nearest neighbor spin model with *positive* weights given as follows. Let $S_j = (h_j, \sigma_j)$ denote spin vector with h_j and σ_j taking values in a hyperboloid and the sphere S^2 respectively. The Gibbs weight is then proportional to

$$\prod_{j=0}^{L} (h_j \cdot h_{j+1} + \sigma_j \cdot \sigma_{j+1})\, e^{\beta(\sigma_j \cdot \sigma_{j+1} - h_j \cdot h_{j+1})} \ . \tag{4.8}$$

As in classical statistical mechanics, the parameter $\beta > 0$ is referred to as the inverse temperature. The hyperbolic components $h_j = (x_j, y_j, z_j)$ satisfy the constraint $z_j^2 - x_j^2 - y_j^2 = 1$. The Heisenberg spins σ are in $\mathbb{R}^3$ with $\sigma \cdot \sigma = 1$ and the dot product is Euclidean. On the other hand the dot product for the h spins is hyperbolic: $h \cdot h' = zz' - xx' - yy'$. It is very convenient to parameterize this hyperboloid with horospherical coordinates $s, t \in \mathbb{R}$:

$$z = \cosh t + s^2 e^t/2, \quad x = \sinh t - s^2 e^t/2, \quad y = se^t. \tag{4.9}$$

The reader can easily check that $z_j^2 - x_j^2 - y_j^2 = 1$ is satisfied for all values of s and t. This parametrization plays an important role in later analysis of hyperbolic sigma models [14, 47]. The integration measure in σ is the uniform measure over the sphere and the measure over h_j has the density $\prod e^{t_j} ds_j\, dt_j$. At the end points of the chain we have set $s_0 = s_L = t_0 = t_L = 0$. Thus we have nearest neighbor hyperbolic spins (Boson–Boson sector) and Heisenberg spins (Fermion–Fermion sector) coupled via the Fermion–Boson determinant. In general, this coupling quite complicated. However in 1D it is given by $\prod_j (h_j \cdot h_{j+1} + \sigma_j \cdot \sigma_{j+1})$. As in (4.5), the inverse temperature $\beta \approx W^2 \rho(E)^2$ where $\rho(E)$ is the density of states, and W is the band width. When $\beta \gg 1$, (4.8) shows that the spins tend to align over long distances.

By adapting the recent work of [12] it can be proved that the model given by (4.8) has a localization length proportional to β for all large β. More precisely it is shown that the conductance goes to zero exponentially fast in L, for $L \gg \beta$.

4.1.3 SUSY Sigma Models in 3D

Although the SUSY sigma models are widely used in physics to make detailed predictions about eigenfunctions, energy spacings and quantum transport, for disordered quantum systems, there is as yet no rigorous analysis of the $SU(1,1|2)$ Efetov models in 2 or more dimensions. Even in one dimension, where the problem can be reduced to a transfer matrix, rigorous results are restricted to the sigma model mentioned above. A key difficulty arises from the fact that SUSY lattice field models cannot usually be treated using probabilistic methods due to the presence of Grassmann variables and strongly oscillatory factors.

In recent work with Disertori and Zirnbauer [12, 14] we have established a phase transition for a *simpler* SUSY hyperbolic sigma model in three dimensions. We shall call this model the $H^{2|2}$ model. The notation refers to the fact that the field takes values in hyperbolic 2 space augmented with 2 Grassmann variables to make it supersymmetric. This model, introduced by Zirnbauer [52], is expected to reflect the qualitative behavior, such as localization or diffusion, of random band matrices in any dimension. The great advantage of the $H^{2|2}$ model is that the Grassmann variables can be traced out producing a statistical mechanics model with *positive* but nonlocal weights. (The non locality arises from a determinant produced by integrating out the Grassmann fields.) This means that probabilistic tools can be applied. In fact we shall see that quantum localization and diffusion is closely related to *a random walk in a random environment.* This environment is highly correlated and has strong fluctuations in 1 and 2D.

In Sect. 4.8 we will describe the $H^{2|2}$ model and sketch a proof of a phase transition as $\beta(E) > 0$ goes from small to large values. Small values of β, high temperature, will correspond to localization—exponential decay of correlations and lack of conductance. In three dimensions, we shall see that large values of β correspond to quantum diffusion and extended states. Standard techniques for proving the existence of phase transitions, such as reflection positivity, do not seem to apply in this case. Instead, the proof of this transition relies crucially on Ward identities arising from SUSY symmetries of the model combined with estimates on random walk in a random environment. The simplest expression of these Ward identities is reflected by the fact that the partition function is identically one for all parameter values which appear in a SUSY fashion. Thus derivatives in these parameters vanish and yield identities. The SUSY $H^{2|2}$ model is nevertheless highly non trivial because physical observables are not SUSY and produce interesting and complicated correlations. Classical Ward identities will be discussed in Sects. 4.6 and 4.7. In Sect. 4.9 we give a very brief description of Efetov's sigma model.

4.2 Classical Models of Quantum Dynamics

In this section we digress to describe two classical models which have some connection with quantum dynamics. The first is called linearly edge reinforced random walk, ERRW. This history dependent walk favors moving along edges it has visited in the past. Diaconis [10] introduced this model and discovered that ERRW can be expressed as random walk in a highly correlated random environment. The statistical description of the random environment appears to be closely related but not identical to that of the SUSY, $H^{2|2}$ model described in Sect. 4.8. In both cases it is important to observe that the generator for the walk is not uniformly elliptic, making it possible to get non-diffusive dynamics. The second model, called the Manhattan model, arises from network models developed to understand the Quantum Hall transition. This model is equivalent to the behavior of a particle undergoing a random unitary evolution. The remarkable feature of this model is that after the randomness is integrated out, the complex phases along paths are canceled. Thus paths have positive weights and the motion has a probabilistic interpretation.

To define linearly edge reinforced random walk (ERRW), consider a discrete time walk on $\mathbb{Z}^d$ starting at the origin and let $n(e,t)$ denote the number of times the walk has visited the edge e up to time t. Then the probability $P(v, v', t+1)$ that the walk at vertex v will visit a neighboring edge $e = (v, v')$ at time $t+1$ is given by

$$P(v, v', t+1) = (\beta + n(e,t))/S_\beta(v,t) \tag{4.10}$$

where S_β is the sum of $\beta + n(e', t)$ over all the edges e' touching v. The parameter β is analogous to β in (4.8) or to the β in the $\mathbb{H}^{2|2}$ model defined later. Note that if β is large, the reinforcement is weak and in the limit $\beta \to \infty$, (4.10) describes simple random walk.

In 1D and 1D strips, ERRW is *localized* for any value of $\beta > 0$ [35]. This means that the probability of finding an ERRW, $w(t)$, at a distance r from its initial position is exponentially small in r, thus

$$Prob\,[|w(t) - w(0)| \geq r] \leq Ce^{-r/\ell}. \tag{4.11}$$

Merkl and Rolles [34] established this result by proving that conductance across an edge goes to zero exponentially fast with the distance of the edge to the origin. Note that in this model, (and in the $H^{(2|2)}$ model with $\epsilon_j = \delta_{j,0}$), the random environment is not translation invariant and the distribution of the local conductance depends on starting point of the walk. Localization is proved, using a variant of the Mermin–Wagner type argument and the localization length ℓ is proportional to $\beta|S|$ where $|S|$ is the width of the strip. Using similar methods, Merkl and Rolles also show that in 2D the local conductance goes to zero slowly away from 0. In 3D, there are no rigorous theorems for ERRW. Localization probably occurs for strong reinforcement, i.e., for β small. It is natural to conjecture that in 2D, ERRW is always exponentially localized for all values of reinforcement. On the Bethe lattice,

Fig. 4.1 Manhattan lattice

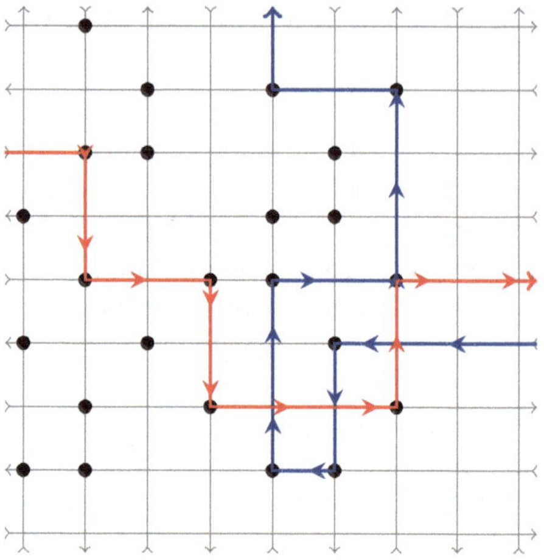

Pemantle [38] proved that ERRW has a phase transition. For $\beta \gg 1$ the walk is transient whereas for $0 < \beta \ll 1$ the walk is recurrent. See [34, 38] for reviews of this subject. It is unknown whether ERRW has a phase transition in 3D.

4.2.1 Manhattan Model

Another classical model which is closely related to quantum dynamics, is defined on the oriented Manhattan lattice. This model was analyzed by Beamond et al. [3] following related work of Gruzberg et al. [30]. In this model disorder occurs by placing obstructions at each vertex, independently with probability $0 < p < 1$. A particle moves straight along the oriented lattice lines according to the orientation of the edges until it meets an obstruction. The particle must turn at obstruction in the direction of the orientation. Note that orbits can traverse an edge at most once. See Fig. 4.1. This model is closely related to a disordered quantum network model (class C). It can also be expressed as a history dependent walk. If $p > 1/2$ all paths form closed loops with finite expected length. This follows (J. Chalker) by comparing the model to classical bond percolation. For p small (weak disorder) the particle is expected to exhibit diffusion for long time scales. Nevertheless, the renormalization group analysis of [3], indicates that for all $p > 0$, every path of this walk is closed with probability one and has a *finite expected diameter*. The average diameter is predicted to be huge $\sim \exp Cp^{-2}$ when p is small. This prediction is consistent with those for the crossover from diffusion to Anderson localization in two dimensions. At a mathematical level, little is known when $0 < p < 1/2$.

Note that the Manhattan model is quite different from the Ruijgrok–Cohen mirror model on the unoriented square lattice. In this case one randomly places mirrors with probability p at the vertices, angled at ±45° with respect to the x axis. Orbits are obtained by imagining that a light beam light travels along lattice lines until it is reflected by a mirror. If $p = 1$ and the choice of angles is independent with probability $1/2$, the model is equivalent to *critical bond percolation*. Although the loops have finite length with probability one, their average length is *infinite*. For $0 < p < 1$ little is known rigorously but one expects that the average loop length to still be infinite.

Remark. Although the two models described above are quite special, they share one of the essential difficulties of the random Schrödinger models and RBM—they are highly non-Markovian. It is this feature that makes these models challenging to analyze. On the other hand, it is the memory or non-Markovian character which is partly responsible for localization. The problem of proving strictly subballistic dynamics, $R^2(t) \propto t^{2\alpha}$ with $\alpha < 1$, is also unresolved for these classical systems in dimension two or more.

4.3 Introduction Random Matrix Ensembles and SUSY

In this section we shall define various Gaussian matrix ensembles. The simplest of these is GUE—Gaussian Unitary ensemble. In this case the matrix entries H_{ij} are mean zero, complex, independent random variables for $i \le j$ and $1 \le i, j \le N$. Since H has a Gaussian distribution it suffices to specify its covariance:

$$<H_{ij}\bar{H}_{i'j'}> = <H_{ij}H_{j'i'}> = \delta(ii')\delta(jj')/N. \tag{4.12}$$

The average over the randomness or disorder is denoted by $< \cdot >$ and $\bar{H}$ denotes the complex conjugate of H. The density for this ensemble is given by

$$1/Z_N \; e^{-N\,tr\,H^2/2} \prod dH_{ii} \prod_{i<j} dH_{ij}^{Re} dH_{ij}^{Im}.$$

The factor of $1/Z_N$ ensures that the integral is 1. It is clear that H and U^*HU have identical distributions for any fixed unitary matrix U. This invariance is a crucial feature in the classical analysis of such matrices via orthogonal polynomials. It also holds when $tr\,H^2$ above is replaced by $tr\,V(H)$ for polynomials V, which are bounded from below [8, 9, 37]. However, Wigner matrices, and random band matrices do not have unitarily invariant distributions and new methods are needed to obtain the desired spectral information. See [19, 46] for recent reviews of Wigner matrices and random band matrices.

To define random band matrices, RBM, with a Gaussian distribution, let i and j range over a periodic box $\Lambda \subset \mathbb{Z}^d$ of volume $|\Lambda|$ and set

$$<H_{ij}\bar{H}_{i'j'}> = <H_{ij}H_{j'i'}> = \delta(ii')\delta(jj')J_{ij}. \tag{4.13}$$

Here J_{ij} is a symmetric function which is *small* for large $|i - j|$. We shall assume that for fixed i, $\sum_j J_{ij} = 1$. With this normalization, it is known that most of the spectrum of H is concentrated in the interval $[-2, 2]$ with high probability. One especially convenient choice of J is given by the lattice Green's function

$$J_{jk} = (-W^2\Delta + 1)^{-1}(j,k) \tag{4.14}$$

where Δ is the lattice Laplacian on Λ with periodic boundary conditions

$$\Delta f(j) = \sum_{|j'-j|=1} (f(j') - f(j)).$$

Note that with this choice of J, $\sum_j J_{ij} = 1$ and the variance of the matrix elements is exponentially small when $|i - j| \gg W$. In fact, in one dimension $J_{ij} \approx e^{-|i-j|/W}/W$. Here $|i - j|$ is the natural distance in the periodic box. Hence W will be referred to as the width of the band. The advantage of (4.14) is that its *inverse*, which appears by duality in certain SUSY formulations, has only *nearest neighbor* matrix elements and W^2 is just a parameter, see (4.31) below.

The density for the Gaussian orthogonal ensemble, GOE, is also proportional to $exp(-N\, trH^2/2)$ but the matrix H is real symmetric. The covariance is given by $< H_{jj}^2 >= 1/N$ and

$$< H_{ij}H_{kl} >= (2N)^{-1}\{\delta(ik)\delta(jl) + \delta(il)\delta(jk)\}, \quad i \neq j.$$

The symmetric RBM are defined by introducing J_{ij} as in (4.13). Although the SUSY formalism can also be applied to the symmetric case, the formulas are a bit more complicated and we shall not discuss them.

4.3.1 Some Conjectures About RBM and Random Schrödinger

Let us now compare lattice random Schrödinger matrices (RS) on $\mathbb{Z}^d$ given by

$$H_{RS} = -\Delta + \lambda v_j$$

and RBM (4.13) of width W on $\mathbb{Z}^d$. Above, v_j are assumed to be independent identically distributed Gaussian random variables of mean 0 and variance $< v_j^2 > = 1$. The parameter $\lambda > 0$ measures the strength of the disorder. See [22, 45] for recent reviews of mathematical aspects of Anderson localization. Although RBM and RS look quite different, they are both local, that is their matrix elements j, k are small (or zero) if $|j - k|$ is large. The models are expected to have the same qualitative properties when $\lambda \approx W^{-1}$. For example, eigenvectors for RS are known to decay exponentially fast in one dimension with a localization length proportional to λ^{-2}.

On the other hand for 1D, RBM the localization length is known to be less than W^8 by work of Schenker [42], and is expected to be approximately W^2. From the point of view of perturbation theory, if we compute the bare diffusion constant D_0, obtained by summing ladder diagrams, then we have $D_0 \propto \lambda^{-2}$, W^2 for the RS and RBM respectively. See Appendix 4.11 for a review of perturbation theory for RS and RBM.

Localization is reasonably well understood for RS in one dimension for any disorder. Localization has also been proved on $\mathbb{Z}^d$ for strong disorder or at energies where the density of states $\rho(E)$ is very small. See [45] for references to the mathematical literature. On the other hand a mathematical proof of quantum diffusion in 3D, as defined by (4.3) as $\epsilon \downarrow 0$ for RS or RBM for fixed W, remains a major open problem. In 2D it is conjectured that all states are localized with a localization length growing exponentially fast in W^2 or λ^{-2} for real symmetric matrices. Note the similarity of this conjecture with that for the Manhattan model defined in Sect. 4.2.

4.3.2 *Conjectured Universality of GUE and GOE Models*

GUE or GOE matrices are mean field and have no spatial or geometric structure. Nevertheless, the local eigenvalue statistics of these models are expected to match those of certain RBM for large N. For example, in 1D the RBM of the form (4.13), (4.14) should have the same local energy level statistics as GUE as $N \to \infty$, (modulo a simple rescaling), provided that $W^2 \gg N$ and energies lie in the bulk, $|E| \leq 2 - \delta$. In 3D universality is expected to hold for E in the bulk, and for fixed $W >> 1$ independent of $N = |\Lambda| \to \infty$. A. Sodin [43] recently proved universality of local energy spacing about the spectral edge, $E \approx 2$ provided $W \gg N^{5/6}$. His result is sharp in the sense that for smaller values of W another distribution appears. Universality has been established for Wigner matrices at all energies [19].

In terms of SUSY statistical mechanics, universality of local energy correlations for Hermitian band matrices, can be intuitively understood as follows. For appropriate values of E, W and dimension, energy correlations are essentially governed by the saddle manifold which is the orbit of a single saddle point under the action of $SU(1,1|2)$. This manifold only depends on the symmetry of the ensemble. See Sect. 4.3 of [36] for a more detailed (but not rigorous) explanation of universality and explicit corrections to it arising from finite volume effects. Universality fails when the fluctuations about the saddle manifold are too pronounced. This can happen if W is not large enough or when $d \leq 2$. In such cases localization may occur and the eigenvalue spacing will be Poisson. Note that in 1D localization effects should not appear unless $N \gg W^2$.

The local eigenvalue spacing statistics for a random Schrödinger matrix in a large box is expected to match those of GOE after simple rescaling, provided that one looks in a region of energies where diffusion should appear, i.e., (4.3) holds. RS corresponds to GOE rather than GUE because it is real symmetric matrix and the corresponding saddle manifold is different from GUE. See [48] for some interesting results showing that in some special, limiting cases the local eigenvalue spacings for the random Schrödinger matrix has GOE statistics.

Remark. The local eigenvalue statistics of GUE appear to be mysteriously connected to the statistics of the spacings of zeros of the Riemann zeta function. See [39] and Keating's article in [6] for reviews of this conjecture and others concerning the relation of random matrix theory and number theory.

4.3.3 Green's Functions and Gaussian Integrals

For an $N \times N$ Hermitian matrix H, define the inverse matrix:

$$G(E_\epsilon) = (E_\epsilon - H)^{-1} \qquad where \qquad E_\epsilon = E - i\epsilon. \tag{4.15}$$

This a bounded matrix provided that E is real and $\epsilon > 0$. The Green's matrix is denoted $G(E_\epsilon)$, and $G(E_\epsilon; k, j)$, its matrix elements, will be called the Green's function.

Let $z = (z_1, z_2, \ldots, z_N)$ with $z_j = x_j + iy_j$ denote an element of $\mathbb{C}^N$ and define the quadratic form

$$[z;\, H\, z] = \sum_{i,j} \bar{z}_k\, H_{kj}\, z_j. \tag{4.16}$$

Then we can calculate the following Gaussian integrals:

$$\int e^{-i[z;(E_\epsilon - H)z]} Dz = (-i)^N \det(E_\epsilon - H)^{-1} \;\; where \;\; Dz \equiv \prod_j^N dx_j dy_j/\pi \tag{4.17}$$

and

$$\int e^{-i\,[z;(E_\epsilon - H)z]} z_k \bar{z}_j\; Dz = (-i)^{N+1} \det(E_\epsilon - H)^{-1}\, G(E_\epsilon; k, j). \tag{4.18}$$

It is important to note that the integrals above are *convergent* provided that $\epsilon > 0$. The quadratic form $[z;\,(E - H)z]$ is real so its contribution only oscillates. The factor of $i = \sqrt{-1}$ in the exponent is needed because the matrix $E - H$ has an indefinite signature when E is in the spectrum of H.

There is a similar identity in which the complex commuting variables z are replaced by anticommuting Grassmann variables $\psi_j, \bar{\psi}_j$, j $= 1, 2 \ldots$N. Let A be an $N \times N$ matrix and as in (4.16) define

$$[\psi; A\, \psi] = \sum \bar{\psi}_k\, A_{kj}\, \psi_j.$$

Then we have

$$\int e^{-[\psi;A\psi]} D\psi = \det A \quad where \quad D\psi \equiv \prod_j^N d\bar{\psi}_j d\psi_j. \tag{4.19}$$

See Appendix 4.10 for a brief review of Gaussian and Grassmann integration. The Grassmann integral is introduced so that we can cancel the unwanted determinant in (4.18) and so that we can perform the average over the randomness in H. Thus we obtain a SUSY representation for the Green's function:

$$G(E_\epsilon; k, j) = i \int e^{-i[z;(E_\epsilon - H)z]} e^{-i[\psi;(E_\epsilon - H)\psi]} z_k \bar{z}_j \, Dz \, D\psi. \tag{4.20}$$

Equation (4.20) is the starting point for all SUSY formulas. Notice that if H has a Gaussian distribution, the expectation of (4.20) can be explicitly performed since H appears linearly. We obtain:

$$< G(E_\epsilon; k, j) >= i \int e^{-iE_\epsilon([z;z]+[\psi;\psi])} e^{-\frac{1}{2}<\{[z;Hz]+[\psi;H\psi]\}^2>} z_k \bar{z}_j \, Dz \, D\psi. \tag{4.21}$$

The resulting lattice field model will be quartic in the z and ψ fields. Notice that if the observable $i\, z_k\, \bar{z}_j$ were absent from (4.21), then the determinants would cancel and the integral would be equal to 1 for all parameters. Thus in SUSY systems, the *partition function is identically 1.*

In a similar fashion we can obtain more complicated formulas for

$$\Gamma(E_\epsilon, j) =< G(E_\epsilon; 0, j) \bar{G}(E_\epsilon; 0, j) > .$$

To do this we must introduce additional variables $w \in \mathbb{C}^N$ and independent Grassmann variables $\chi, \bar{\chi}$ to obtain the second factor, $\bar{G}$, which is the complex conjugate of G. See Sects. 4.7 and 4.9 below.

4.4 Averaging $Det(E_\epsilon - H)^{-1}$

Before working with Grassmann variables we shall first show how to calculate the average of the inverse determinant over the Gaussian disorder using only the complex variables z_j. Although this average has no physical significance, it will illustrate some basic mathematical ideas.

First consider the simplest case: H is an $N \times N$, GUE matrix. Let us apply (4.17) to calculate

$$(i)^{-N} < \det(E_\epsilon - H)^{-1} >=< \int e^{-i[z;\,(E_\epsilon - H)z]} \, Dz > . \tag{4.22}$$

Since the expectation is Gaussian and the covariance is given by (4.12), the average over H can be expressed as follows:

$$< e^{-i[z;\,Hz]} >_{GUE} = e^{-1/2<[z;\,Hz]^2>} = e^{-\frac{1}{2N}[z;z]^2}. \qquad (4.23)$$

The most direct way to estimate the remaining integral over the z variables is to introduce a new coordinate $r = [z, z] = \sum |z_j|^2$. Then we have

$$< \det(E_\epsilon - H)^{-1} >= C_N \int_0^\infty e^{-\frac{1}{2N}r^2 - iE_\epsilon r} r^{N-1}\, dr$$

where C_N is an explicit constant related to the volume of the sphere in 2N dimensions. It is convenient to rescale $r \to Nr$ and obtain an integral of the form

$$\int_0^\infty e^{-N(r^2/2 - \ln r - iE_\epsilon r)} dr/r.$$

The method of steepest descent can now be applied. We deform the contour of integration over r so that it passes through the saddle point. The saddle point r_s is obtained by setting the derivative of the exponent to 0: $r_s - 1/r_s - iE_\epsilon = 0$ This is a quadratic equation with a solution $r_s = iE/2 + \sqrt{1-(E/2)^2}$. The contour must be chosen so that the absolute value of the integrand is dominated by the saddle point.

Exercise. Stirling's formula:

$$N! = \int_0^\infty e^{-t} t^N dt \approx N^N e^{-N} \sqrt{2N\pi}.$$

To derive this let $t = Ns$ and expand to quadratic order about the saddle point s = 1. The square root arises from the identity $N \int e^{-Ns^2/2} ds = \sqrt{2N\pi}$. We shall see many SUSY calculations are related to generalizations of the gamma function.

Remark. For other ensembles, radial coordinates do not suffice and one must compute the Jacobian for the new collective variables. See the section on matrix polar coordinates in Appendix 4.10. Collective coordinates can also be set up with the help of the coarea formula.

An alternate way to compute $< \det(E_\epsilon - H)^{-1} >$ uses the Hubbard–Stratonovich transform. In its simplest form, introduce a real auxiliary variable a to unravel the quartic expression in z as follows:

$$e^{-\frac{1}{2N}[z;\,z]^2} = \sqrt{N/2\pi} \int e^{-Na^2/2} e^{ia[z;\,z]} da. \qquad (4.24)$$

If we apply (4.24) to (4.22) and (4.23), the z variables now appear quadratically and we can integrate over them. Since there are N independent z_j to integrate, we get a factor $(E_\epsilon - a)^{-N}$, hence:

$$< \det(E_\epsilon - H)^{-1} >= i^N \sqrt{N/2\pi} \int e^{-Na^2/2} (E_\epsilon - a)^{-N} da. \qquad (4.25)$$

Let

$$f(a) = N\,[a^2/2 + ln(E_\epsilon - a)].$$

The saddle point a_s is obtained by setting $f'(a_s) = 0$. This gives a quadratic equation whose solution is

$$a_s = E_\epsilon/2 + i\sqrt{1 - (E_\epsilon/2)^2}. \qquad (4.26)$$

We shift our contour of integration by $a \to a + a_s$. Note $|a_s| \approx 1$ and that we have chosen the $+$ sign in (4.26) so that the pole of $(E - i\epsilon - a)^{-N}$ has not been crossed. Along this contour one checks that for E satisfying $|E| \leq 1.8$, the maximum modulus of the integrand occurs at the saddle a_s. In particular, this deformation of contour avoids the small denominator $E_\epsilon - a$ occurring when $a \approx E$. For energies such that $1.8 \leq |E| \leq 2$ another contour must be used [11]. Note that the Hessian at the saddle is

$$f''(a_s) = N(1 - a_s^2) = N\{2(1 - (E/2)^2) - iE\sqrt{1 - (E/2)^2}\} \qquad (4.27)$$

has a positive real part for $|E| < 2$. To complete our estimate of (4.25) for large N we calculate the quadratic correction to the saddle point:

$$< \det(E_\epsilon - H)^{-1} >\approx i^N \sqrt{N/2\pi} e^{-Nf(a_s)} \int e^{-N/2 f''(a_s)a^2} da.$$

Higher order corrections to the saddle point are suppressed by powers of N^{-1}.

Remark. The imaginary part of right hand side of (4.25), (N even), is proportional to

$$e^{-E^2/2} H_N(E\sqrt{N/2})$$

where H_N is the N'th Hermite polynomial. This can be seen by integrating by parts $N - 1$ times. Thus such identities point to a connection between the method of orthogonal polynomials and SUSY statistical mechanics. See [11] for more details.

4.4.1 Gaussian Band Matrices and Statistical Mechanics

Now let us consider the more general case when H is a finite Gaussian random band matrix indexed by lattice sites in a periodic box $\Lambda \subset \mathbb{Z}^d$ with covariance given by (4.13) and (4.14): $J_{jk} = (-W^2\Delta + 1)^{-1}(j,k)$. Then we have

$$< e^{-i[z;Hz]} >= e^{-1/2<[z;Hz]^2>} = e^{-1/2\sum |z_i|^2 J_{ij} |z_j|^2}. \tag{4.28}$$

In order to average over the z variables we introduce real Gaussian auxiliary fields a_j with 0 mean and covariance J_{ij}. Let $< \cdot >_J$ be the corresponding expectation so that $< a_i a_j >_J = J_{ij}$ and

$$e^{-1/2\sum_{ij} |z_i|^2 J_{ij} |z_j|^2} =< e^{i\sum a_j |z_j|^2} >_J . \tag{4.29}$$

This is again the Hubbard–Stratonovich trick. We can now average over the z's since they appear quadratically. By combining (4.22), (4.28) and (4.29)

$$\begin{aligned} < \det(E_\epsilon - H)^{-1} > &=< \int e^{-i[z;(E_\epsilon - H)z]} Dz > \qquad (4.30) \\ &=< \int e^{-i[z;(E_\epsilon - a)z]} Dz >_J =< \prod_{j\in\Lambda} (E_\epsilon - a_j)^{-1} >_J . \end{aligned}$$

Since by (4.14), the Gaussian a_j fields have covariance $(-W^2\Delta + 1)^{-1}$, it follows that (4.30) is proportional to:

$$\int e^{-\frac{1}{2}\sum_j (W^2(\nabla a)_j^2 + a_j^2)} \prod_{j\in\Lambda} (E_\epsilon - a_j)^{-1}\, da_j . \tag{4.31}$$

The expression $(\nabla a)_j^2$ is the square of finite difference gradient at j. The large parameter W tends to make the a_j fields constant over a large range of lattice sites. The saddle point can be calculated as it was for GUE and we find that it is independent of the lattice site j and is again given by (4.26). We deform the contour of integration $a_j \to a_j + a_s$. We must restrict $|E| \le 1.8$ as in the case of GUE so that the norm of the integrand along the shifted contour is maximized at the saddle. The Hessian at the saddle is given by

$$H_s'' = -W^2\Delta + 1 - a_s^2. \tag{4.32}$$

Since $Re(1 - a_s^2) > 0$, $[H'']^{-1}(j,k)$ decays exponentially fast for separations $|j - k| \gg W$.

Since (4.31) is an integral over many variables a_j, $j \in \Lambda$, its behavior is not so easy to estimate as (4.25). To control fluctuations away from the saddle one uses the exponential decay of $[H'']^{-1}$ and techniques of *statistical mechanics* such as the cluster or polymer expansion [40], about blocks of side W. This will show that the integral can be approximated by a product of nearly independent integrals over boxes of size W. It enables one to justify the asymptotic expansion for

$$|\Lambda|^{-1} \log < \det(E_\epsilon - H)^{-1} >$$

in powers of W^{-1}. Moreover, one can prove that with respect to the expectation defined using (4.31), a_j, a_k are exponentially independent for $|j - k| \gg W$ uniformly in $\epsilon > 0$. The main reason this works is that the inverse of the Hessian at a_s has exponential decay and the corrections to the Hessian contribution are small when W is large. As in (4.25), corrections beyond quadratic order are small for large N. See [7, 13] for the SUSY version of these statements.

Remark. In one dimension, (4.31) shows that we can calculate the $< \det(E_\epsilon - H)^{-1} >$ using a nearest neighbor transfer matrix. In particular if the lattice sites j belong to a circular chain of length L, there is an integral operator $T = T(a, a')$ such that $< \det(E_\epsilon - H)^{-1} >= Tr\, T^L$.

Remark. For the lattice random Schrödinger matrix, $H = -\Delta + \lambda v$, in a finite box, with a Gaussian potential v of mean 0 and variance 1, it is easy to show that

$$< \det(E_\epsilon - H)^{-1} >= \int e^{-i\,[z\,;(E_\epsilon+\Delta)z]-\frac{\lambda^2}{2}\sum_j |\bar{z}_j z_j|^2} Dz.$$

The complex variables z_j are indexed by vertices in a finite box contained in $\mathbb{Z}^d$. For small λ this is a highly oscillatory integral which is more difficult than (4.31) to analyze. Rigorous methods for controlling such integrals are still missing when λ is small and $d \geq 2$. Note that for RBM, the oscillations are much less pronounced because the off diagonal terms are also random. In this case the factor

$$\exp -\frac{1}{2}(\sum_j W^2(\nabla a)_j^2 + a_j^2)$$

dominates our integral for large W.

4.5 The Density of States for GUE and RBM

The *integrated* density of states for an $N \times N$ Hermitian matrix H is denoted $n(E) = \int^E d\rho(E')$ is the fraction of eigenvalues less than E and $\rho(E)$ denotes the density of states. The average of the density of states is given by the expression

$$< \rho_\epsilon(E) >= \frac{1}{N} tr < \delta_\epsilon(H - E) >= \frac{1}{N\pi} tr\, Im \; < G(E_\epsilon) > \tag{4.33}$$

as $\epsilon \downarrow 0$. Here we are using the well known fact that

$$\delta_\epsilon(x - E) \equiv \frac{1}{\pi}\frac{\epsilon}{(x - E)^2 + \epsilon^2} = \frac{1}{\pi} Im(E_\epsilon - x)^{-1}$$

is an approximate delta function at E as $\epsilon \to 0$.

Remarks. The Wigner semicircle distribution asserts that the density of states of a GUE matrix is given by $\pi^{-1}\sqrt{1-(E/2)^2}$. Such results can be proved for many ensembles including RBM by *first fixing* ϵ in (4.33) and then letting N*, and* $W \to \infty$. Appendix 4.11 gives a simple formal derivation of this law using self-consistent perturbation theory. Note that the parameter ϵ is the scale at which we can resolve different energies. For a system of size N we would like to understand the number of eigenvalues in an energy window of width $N^{-\alpha}$ with $\alpha \approx 1$. Thus we would like to get estimates for $\epsilon \approx N^{-\alpha}$. Such estimates are difficult to obtain when $\alpha \approx 1$ but have recently been established for Wigner ensembles, see [20]. Using SUSY [7, 13] estimates on the density of states for a special class of Gaussian band matrices indexed by $\mathbb{Z}^3$ are proved *uniformly* in ϵ and the size of the box for fixed $W \geq W_0$. See the theorem in this section.

Note that the density of states is *not* a universal quantity. At the end of Appendix 4.11 we show that 2 coupled GUE $N \times N$ matrices has a density of states given by a cubic rather than quadratic equation.

We now present an identity for the average Green's function for GUE starting from (4.21). Note that by (4.12)

$$-\frac{1}{2} < ([z; Hz] + [\psi; H\psi])^2 >_{GUE}$$
$$= -\frac{1}{2N}\{[z;z]^2 - [\psi;\psi]^2 - 2[\psi;z][z;\psi]\}. \tag{4.34}$$

Let us introduce another real auxiliary variable $b \in \mathbb{R}$ and apply the Hubbard–Stratonovich transform to decouple the Grassmann variables. As in (4.24) we use the identity

$$e^{[\psi;\psi]^2/2N} = \sqrt{N/2\pi} \int db\, e^{-Nb^2/2}\, e^{-b[\psi;\psi]}. \tag{4.35}$$

Now by combining (4.24) and (4.35) we see that the expressions are quadratic in z and ψ. Note that

$$\int e^{-\sum^N \bar{\psi}_j \psi_j (b+iE_\epsilon)}\, D\psi\, Dz = i^N\,(E_\epsilon - ib)^N.$$

The exponential series for the cross terms appearing in (4.34), $[\psi; z][z; \psi]$, terminates after one step because $[\psi; z]^2 = 0$. We compute its expectation with respect to the normalized Gaussian measure in ψ, z and get

$$< [\psi;z][z;\psi] > = < \sum_{ij} z_i \bar{\psi}_i \bar{z}_j \psi_j > = \sum_j < \bar{\psi}_j \psi_j \bar{z}_j z_j >$$
$$= -N(E_\epsilon - ib)^{-1}(E_\epsilon - a)^{-1}.$$

Thus after integrating over both the z and ψ fields and obtain the expression:

$$
\begin{aligned}
<\frac{1}{N} trG(E_\epsilon)> &= N^{-1}\int [z;z]\, e^{-\frac{1}{2N}\{[z;z]^2-[\psi;\psi]^2-2[\psi;z][z;\psi]\}}\, Dz\, D\psi \\
&= N/2\pi \int dadb\,(E_\epsilon-a)^{-1}e^{-N(a^2+b^2)/2}(E_\epsilon-ib)^N(E_\epsilon-a)^{-N} \\
&\times\left[1-\frac{N+1}{N}(E_\epsilon-a)^{-1}(E_\epsilon-ib)^{-1}\right]\approx<(E_\epsilon-a)^{-1}>_{SUSY}.
\end{aligned}
\tag{4.36}
$$

The first factor of $(E_\epsilon - a)^{-1}$ on the right hand side roughly corresponds to the trace of the Green's function. The partition function is

$$
N/2\pi\int dadb\, e^{-N(a^2+b^2)/2}(E_\epsilon-ib)^N(E_\epsilon-a)^{-N}\,[1-(E_\epsilon-a)^{-1}(E_\epsilon-i\,b)^{-1}] \equiv 1 \tag{4.37}
$$

for all values of E, ϵ and N. This is due to the fact that if there is no observable the determinants cancel. The last factor in (4.37) arises from the cross terms and is called the Fermion–Boson *(FB)* contribution. It represents the coupling between the determinants. For band matrices it is useful to introduce auxiliary dual Grassmann variables (analogous to a and b) to treat this contribution. See [11, 13, 36].

The study of $\rho(E)$ reduces to the analysis about the saddle points of the integrand. As we have explained earlier, there is precisely one saddle point

$$
a_s(E) = E_\epsilon/2 + i\sqrt{1-(E_\epsilon/2)^2} \tag{4.38}
$$

in the a field. However, the b field has two saddle points

$$
i\,b_s = a_s, \quad and \quad i\,b_s^{'} = E_\epsilon - a_s = \bar{a}_s.
$$

Hence, both saddle points (a_s, b_s) and $(a_s, b_s^{'})$ contribute to (4.36).

Let us briefly analyze the fluctuations about the saddles as we did in (4.26), (4.27). The first saddle gives the Wigner semicircle law. To see this note that the action at a_s, b_s takes the value 1 since $a_s^2 + b_s^2 = 0$ and $(E_\epsilon - i b_s)/(E_\epsilon - a_s) = 1$. The integral of quadratic fluctuations about the saddle,

$$
N\int e^{-N(1-a_s^2)(a^2+b^2)}dadb \tag{4.39}
$$

is exactly canceled by the *FB* contribution at (a_s, b_s). Thus to a high level of accuracy we can simply replace the observable in the SUSY expectation (4.36) by its value at the saddle. This gives us Wigner's semicircle law:

$$\rho(E) = \frac{1}{\pi N} Im\ tr < G(E_\epsilon) > \approx \pi^{-1} Im < (E_\epsilon - a)^{-1} >_{SUSY}$$
$$\approx \pi^{-1} Im(E_\epsilon - a_s)^{-1} = \pi^{-1}\sqrt{1-(E/2)^2}. \qquad (4.40)$$

It is easy to check that the second saddle vanishes when inserted into the *FB* factor because $(E - a_s)(E - ib'_s) = \bar{a}_s a_s = 1$. Thus to leading order it does not contribute to the density of states and hence (4.40) holds. However, the second saddle will contribute a highly oscillatory contribution to the action proportional to

$$\frac{1}{N}\left(\frac{E_\epsilon - a'_s}{E_\epsilon - a_s}\right)^N e^{-N/2(a_s^2 - \bar{a}_s^2)}. \qquad (4.41)$$

If we take derivatives in E, this contribution is no longer suppressed when $\epsilon \approx 1/N$. This result is not easily seen in perturbation theory. I believe this is a compelling example of the nonperturbative power of the SUSY method.

Remarks. If $\epsilon = 0$, then (4.41) has modulus $1/N$. We implicitly assume that the energy E is inside the bulk ie $|E| < 2$. Near the edge of the spectrum the Hessian at the saddle point vanishes and a more delicate analysis is called for. The density of states near $E = \pm 2$ then governed by an Airy function. We refer to Disertori's review of GUE [11], for more details.

4.5.1 Density of States for RBM

We conclude this section with a brief discussion of the average Green's function for RBM in 3D with the covariance J given by (4.14). The SUSY statistical mechanics for RBM is expressed in terms of a_j, b_j and Grassmann fields $\bar{\eta}_j, \eta_j$ with $j \in \Lambda \subset Z^d$ and it has a local (nearest neighbor) weight

$$\exp\Big[-\frac{1}{2}\sum_j \{W^2(\nabla a_j)^2 + W^2(\nabla b_j)^2 + a_j^2 + b_j^2\}\Big]\prod_j \frac{E_\epsilon - ib_j}{E_\epsilon - a_j}$$
$$\times \exp - \sum_j \{W^2 \nabla\bar{\eta}_j \nabla\eta_j + \bar{\eta}_j\eta_j\,(1 - (E_\epsilon - a_j)^{-1}(E_\epsilon - ib_j)^{-1})\}.$$

After the Grassmann variables have been integrated out we get:

$$\exp[-\frac{1}{2}\sum_j \{W^2(\nabla a_j)^2 + W^2(\nabla b_j)^2 + a_j^2 + b_j^2\}]\prod_j \frac{E_\epsilon - ib_j}{E_\epsilon - a_j}$$
$$\times \det\{-W^2\Delta + 1 - \delta_{ij}(E_\epsilon - a_j)^{-1}(E_\epsilon - ib_j)^{-1}\}\, DaDb. \qquad (4.42)$$

Note the similarity of the above weight with (4.31) and (4.37). The determinant is called the Fermion–Boson contribution. In one dimension, the DOS can be reduced to the analysis of a nearest neighbor transfer matrix. Large W keeps the fields a_j, b_j nearly constant. This helps to control fluctuations about saddle point.

Theorem ([13]). Let d = 3, J be given by (4.14) and $|E| \leq 1.8$. For $W \geq W_0$ the average $< G(E_\epsilon, j, j) >$ for RBM is uniformly bounded in ϵ *and* Λ. It is approximately given by the Wigner distribution with corrections of order $1/W^2$. Moreover, we have

$$| \langle G(E_\epsilon; 0, x)\, G(E_\epsilon; x, 0) \rangle | \leq e^{-m|x|} \tag{4.43}$$

for $m \propto W^{-1}$.

The proof of this theorem follows the discussion after (4.31). We deform our contour of integration and prove that for $|E| \leq 1.8$, the dominant contribution comes from the saddle point a_s, b_s which is independent of j. However, as in GUE one must take into account the second saddle which complicates the analysis. Note that the Hessian at the saddle is still given by (4.32). Its inverse governs the decay in (4.43).

Remark. The first rigorous results of this type appear in [7] for the Wegner N orbital model. Their SUSY analysis is similar to that presented in [13].

Remark. As explained at the end of the previous section, one can derive a formula analogous to (4.42) for the random Schrödinger matrix in 2D or 3D, but we do not have good control over the SUSY statistical mechanics for small λ. A formal perturbative analysis of the average Green's function for the random Schrödinger matrix with small λ is given in Appendix 4.11. Note that for large λ, it is relatively easy to control the SUSY integral since in this case the SUSY measure is dominated by the product measure $\prod_j exp[-\lambda^2(\bar{z}_j z_j + \bar{\psi}_j \psi_j)^2]$. As a final remark note that in the product of Green's functions appearing in (4.43), energies lie on the same side of the real axis. The expectation $< |G(E_\epsilon; 0, j)|^2 >$ is quite different and for such cases hyperbolic symmetry emerges, see Sect. 4.7, and gapless modes may appear.

4.6 Statistical Mechanics, Sigma Models, and Goldstone Modes

In this section we shall review some classical sigma models and their lattice field theory counterparts. This may help to see how classical statistical mechanics is related to the SUSY sigma models which will be described later. The symmetry groups described in this section are compact whereas those of related the Efetov models are noncompact as we shall see in Sects. 4.7–4.9.

Consider a lattice model with a field $\phi(j) = (\phi^1(j), \ldots, \phi^{m+1}(j))$ taking values in $\mathbb{R}^{m+1}$ and whose quartic action in given by

$$A_\Lambda(\phi) = \sum_{j \in \Lambda} \{|\nabla\phi(j)|^2 + \lambda(|\phi(j)|^2 - a)^2 + \epsilon \cdot \phi^1\}. \tag{4.44}$$

As before ∇ denotes the nearest neighbor finite difference gradient and Λ is a large periodic box contained in $\mathbb{Z}^d$. The partition function takes the form

$$Z_\Lambda = \int e^{-\beta A_\Lambda(\phi)} \prod_{j \in \Lambda} d\phi_j.$$

To define its sigma model approximation, consider spins, $S_j \in S^m$ where $j \in \Lambda$. More precisely $S_j = (S_j^{(1)}, \ldots S_j^{(m+1)})$ with $S_j{}^2 = S_j \cdot S_j = 1$. For m = 0, the spin takes values ± 1 and this is the Ising spin. The energy or action of a configuration $\{S_j\}$ is given by

$$A_{\bar{\beta},\Lambda}(S) = \bar{\beta} \sum_{j \sim j' \in \Lambda} (S_j - S_{j'})^2 - \epsilon \sum_j S_j^{(1)}. \tag{4.45}$$

The Gibbs weight is proportional to $e^{-A(S)}$, the parameters $\bar{\beta}$, $\epsilon \geq 0$ are proportional to the inverse temperature and the applied field respectively and $j \sim j'$ denote adjacent lattice sites. If $\epsilon > 0$ then the minimum energy configuration is unique and the all the spins point in the direction $(1, 0, \ldots 0)$.

If $\epsilon = 0$, A is invariant under a global rotation of the spins. The energy is minimized when all the spins are pointing in the same direction. This is the ground state of the system. When m = 1 we have $O(2)$ symmetry and for m = 2 the symmetry is $O(3)$ corresponding to the classical rotator (or X–Y) model and classical Heisenberg model respectively. When $\epsilon > 0$ the full symmetry is broken but in the case of $O(3)$ is broken to $O(2)$. The parameter ϵ in (4.45) also appears in SUSY models and is related to the imaginary part of E_ϵ in the Green's function.

Consider the classical scalar $|\phi|^4$ model, (4.44), with $\phi_j \in \mathbb{C}$ with a $U(1)$ symmetry. The idea behind the sigma approximation is that the complex field ϕ_j can be expressed in polar coordinates $r_j\, S_j$ where $|S_j| = 1$. The radial variables r are "fast" variables fluctuating about some effective radius $r*$. If the quartic interaction has the form $\lambda(|\phi_j|^2 - 1)^2$ and $\lambda \gg 1$, $r^* \approx 1$. If we freeze $r_j = r*$ we obtain (4.45) with $\bar{\beta} \approx (r*)^2\beta$. The $|\phi|^4$ model is expected to have the same qualitative properties as the O(2) sigma models. Moreover, at their respective phase transitions, the long distance asymptotics of correlations should be identical. This means that the two models have the same critical exponents although their critical temperatures will typically be different. This is a reflection of the principle universality for second order transitions. Its proof is a major mathematical challenge.

Remark. The role of large λ above will roughly be played by W in the SUSY models for random band matrices.

Now let us discuss properties of the sigma models with Gibbs expectation:

$$< \cdot >_\Lambda (\bar{\beta}, \epsilon) = Z_\Lambda(\bar{\beta}, \epsilon)^{-1} \int \cdot \; e^{-A_{\bar{\beta}\Lambda}(S)} \prod_j d\mu(S_j). \tag{4.46}$$

The measure $d\mu(S_j)$ is the uniform measure on the sphere. The partition function Z is defined so that this defines a probability measure. Basic correlations are the spin–spin correlations $< S_0 \cdot S_x >$ and the magnetization $< S_0^1 >$. Let us assume that we have taken the limit in which $\Lambda \uparrow Z^d$. First consider the case in which here is no applied field $\epsilon = 0$. In one dimension there is no phase transition and the spin–spin correlations decay exponentially fast for all non zero temperatures, (β finite),

$$0 \leq < S_0 \cdot S_x > \leq Ce^{-|x|/\ell} \tag{4.47}$$

where ℓ is the correlation length. For the models with continuous symmetry, $m = 1, 2$, ℓ is proportional to β. However, for the Ising model ℓ is exponentially large in β. At high temperature, β small, it is easy to show that in any dimension the spins are independent at long distances and (4.47) holds.

In dimension $d \geq 2$, the low temperature phase is interesting and more difficult to analyze. In 2D we know that the Ising model has long range order and a *spontaneous magnetization* M for $\beta > \beta_c$. This means that for large x

$$< S_0 \cdot S_x > (\beta, \epsilon = 0) \rightarrow M^2(\beta) > 0 \;\; and \;\; \lim_{\epsilon \downarrow 0} < S_0^{(1)} > (\beta, \epsilon) = M. \tag{4.48}$$

Thus at low temperature spins align even at long distances. Note that the order of limits here is crucial. In (4.48) we have taken the infinite volume limit first and then sent $\epsilon \rightarrow 0$. For any finite box it is easy to show that second the limit is 0 because the symmetry of the spins is present. For this reason (4.48) is called symmetry breaking.

In two dimensions, the Mermin–Wagner Theorem states that a continuous symmetry cannot be broken. Hence, $M = 0$ for the $O(2)$ and $O(3)$ models. In classical XY model with $O(2)$ symmetry there is a Kosterlitz–Thouless transition [25, 32]: the magnetization, M, vanishes but the spin–spin correlation with $\epsilon = 0$ goes from an exponential decay for small β to a slow power law decay $\approx |x|^{-c/\beta}$ for large β.

A major conjecture of mathematical physics, first formulated by Polyakov, states that for the classical Heisenberg model in 2D, the spin–spin correlation always decays exponentially fast with correlation length $\ell \approx e^{\beta}$. This conjecture is related to Anderson localization expected in 2D. It is a remote cousin of confinement for non-Abelian gauge theory. The conjecture, can partly be understood by a renormalization group argument showing that the effective temperature increases under as we scale up distances. The positive curvature of sphere play a key role in this analysis.

In three dimensions it is known through the methods of reflection positivity [27], that there is long range order and continuous symmetry breaking at low temperature. These methods are rather special and do not apply to the SUSY models described here.

Continuous symmetry breaking is accompanied by Goldstone Bosons. This means that there is a very slow decay of correlations. This is most easily seen in the Fourier transform of the pair correlation. Consider a classical XY or Heisenberg model on $\mathbb{Z}^d$. The following theorem is a special case of the Goldstone–Nambu–Mermin–Wagner Theorem. Let M be defined as in (4.48).

Theorem. Let correlations be defined as above and suppose that the spontaneous magnetization $M > 0$. Then

$$\sum_x e^{-ip\cdot x} < S_0^{(2)}\, S_x^{(2)} > (\beta, \epsilon) \geq CM^2(\beta p^2 + \epsilon M)^{-1}. \qquad (4.49)$$

Thus for $p = 0$ the sum diverges as $\epsilon \downarrow 0$. This theorem applies to the case $m = 1, 2$ but not the Ising model. It is also established for lattice field theories with continuous symmetry.

Remark. In 2D, if the spontaneous magnetization were positive, then the integral over p of the right side of (4.49) diverges for small ϵ. On the other hand, the integral over p of the left side equals 1, since $S_0 \cdot S_0 = 1$. Thus M must vanish in 2D. In higher dimensions there is no contradiction since the integral of $|p|^{-2}, |p| \leq \pi$ is finite.

We shall derive a simple Ward identity for the O(2) sigma model in d dimensions. In angular coordinates we have

$$A(\theta) = -\sum_{j\sim j'} \beta \cos(\theta_j - \theta_{j'}) - \epsilon \sum_j cos(\theta_j). \qquad (4.50)$$

Consider the integral in a finite volume: $\int sin(\theta_0) e^{-A(\theta)} \prod_{j\in\Lambda} d\theta_j$. Make a simple change of variables $\theta_j \to \theta_j + \alpha$. The integral is of course independent of α. If we take the derivative in α and then set $\alpha = 0$ the resulting expression must vanish. This yields a simple Ward identity

$$M =< cos(\theta_0) >= \epsilon \sum_j < \sin(\theta_0)\, \sin(\theta_j) > . \qquad (4.51)$$

After dividing by ϵ we obtain the theorem for $p = 0$. Ward identities are often just infinitesimal reflections of symmetries in our system. In principle we can apply this procedure to any one parameter change of variables.

Remark. Note the similarity of (4.51) with the Ward identity (4.4). The density of states $\rho(E)$ plays the role of the spontaneous magnetization M. Moreover, the right side of (4.49) is analogous to (4.3). However, there is an important distinction:

$\rho(E)$ does not vanish in the high temperature or localized regions. Unlike the magnetization in classical models, $\rho(E)$ is *not* an order parameter. Indeed, $\rho(E)$ is not expected to reflect the transition between localized and extended states. So in this respect classical and SUSY models differ. For SUSY or RBM, the localized phase is reflected in a vanishing of the diffusion constant in (4.3), $D(E,\epsilon) \propto \epsilon$. Thus in regions of localization $< |G(E_\epsilon, 0, 0)|^2 >$ diverges as $\epsilon \to 0$ as we see in (4.1).

Proof of Theorem. Let $|\Lambda|$ denote the volume of the periodic box and set $D = |\Lambda|^{-1/2} \sum_j e^{-ip\cdot j} \frac{\partial}{\partial \theta_j}$ and $\hat{S}(p) = |\Lambda|^{-1/2} \sum_j e^{+ip\cdot j} \sin(\theta_j)$. By translation invariance, integration by parts, and the Schwarz inequality we have

$$M =< \cos(\theta_0) >_A =< D\hat{S}(p) >_A =< \hat{S}(p) D(A) >_A \tag{4.52}$$
$$\leq < |\hat{S}(p)|^2 >_A^{1/2} < \overline{D(A)} D(A) >_A^{1/2} .$$

Since $< |\hat{S}(p)|^2 >$ equals the left side of (4.49) the theorem follows by integrating by parts once again,

$$\begin{aligned} < \overline{D(A)} D(A) >_A &= < D \overline{D(A)} >_A \\ &= |\Lambda|^{-1} \sum_{j \sim j'} < 2\beta cos(\theta_j - \theta_{j'})(1 - cos(j - j')p) \\ &\quad + \epsilon cos(\theta_j) >_A \\ &\leq C(\beta p^2 + \epsilon < \cos(\theta_0) >_A) \end{aligned} \tag{4.53}$$

which holds for small p. Here we have followed the exposition in [26].

4.7 Hyperbolic Symmetry

Let us analyze the average of $|\det(E_\epsilon - H)|^{-2}$ with H a GUE matrix. We study this average because it is the simplest way to illustrate the emergence of hyperbolic symmetry. This section analyses the so called Boson–Boson sector in the sigma model approximation. More complicated expressions involving Grassmann variables appear when analyzing the average of $|G(E_\epsilon; j, k)|^2$. This is touched upon in Sect. 4.9.

To begin, let us contrast $< G >$ and $< |G|^2 >$ for the trivial scalar case N = 1. In the first case we see that $\int e^{-H^2}(E_\epsilon - H)^{-1} dH$ is finite as $\epsilon \to 0$ by shifting the contour of integration off the real axis $H \to H + i\delta$ with $\delta > 0$ so that the pole is not crossed. On the other hand, if one considers the average of $|(E_\epsilon - H)|^{-2}$, we cannot deform the contour integral near E and the integral will diverge like ϵ^{-1}. This divergence will be reflected in the hyperbolic symmetry. Later in this section we shall see that in 3D this symmetry will be associated with gapless Goldstone modes. These modes were absent in Sects. 4.3–4.5.

Let $z, w \in C^N$. As in (4.17) we can write:

$$|\det(E_\epsilon - H)|^{-2} = \det(E_\epsilon - H) \times \det(E_{-\epsilon} - H)$$
$$= \int e^{-i\,[z;(E_\epsilon - H)z]}\, Dz \times \int e^{i\,[w;(E_{-\epsilon} - H)w]}\, Dw. \quad (4.54)$$

Note that the two factors are complex conjugates of each other. The factor of i has been reversed in the w integral to guarantee its convergence. This sign change is responsible for the hyperbolic symmetry. The Gaussian average over H is

$$< e^{-i([zHz]-[w,Hw])} >= e^{-1/2<([z,Hz]-[w,Hw])^2>}. \quad (4.55)$$

Note that

$$< ([z,Hz]-[w,Hw])^2 >=< [\sum H_{kj}(\bar{z}_k z_j - \bar{w}_k z_j)]^2 > . \quad (4.56)$$

For GUE the right side is computed using (4.12)

$$< ([z,Hz]-[w,Hw])^2 >= 1/N([z,z]^2 + [w,w]^2 - 2[z,w][w,z]). \quad (4.57)$$

Note that the hyperbolic signature is already evident in (4.55). Following Fyodorov [29], introduce the 2×2 non negative matrix:

$$M(z,w) = \begin{pmatrix} [z,z] & [z,w] \\ [w,z] & [w,w] \end{pmatrix} \quad (4.58)$$

and let

$$L = diag(1,-1). \quad (4.59)$$

Then we see that

$$< |\det(E_\epsilon - H)|^{-2} >= \int e^{-\frac{1}{2N}tr(ML)^2 - iEtr(ML) + \epsilon trM}\, DzDw. \quad (4.60)$$

For a positive 2×2 matrix P, consider the delta function $\delta(P - M(z,w))$ and integrate over z and w. It can be shown [29,47], also see (4.116) in Appendix 4.10, that

$$\int \delta(P - M(z,w))DzDw = (detP)^{N-2}. \quad (4.61)$$

Assuming this holds we can now write the right side in terms of the new collective coordinate P

$$< |\det(E_\epsilon - H)|^{-2} >= C_N \int_{P>0} e^{-\frac{1}{2N}tr(PL)^2} e^{-iEtr(PL) - \epsilon trP} detP^{N-2} dP. \quad (4.62)$$

After rescaling $P \to NP$ we have

$$< |\det(E_\epsilon - H)|^{-2} >= C'_N \int_{P>0} e^{-N\{tr(PL)^2/2+iEtr(PL)+\epsilon trP\}} det P^{N-2}\, dP. \tag{4.63}$$

In order to compute the integral we shall again change variables and integrate over PL. First note that for $P > 0$, PL has two real eigenvalues of opposite sign. This is because PL has the same eigenvalues as $P^{1/2}LP^{1/2}$ which is self adjoint with a negative determinant. Moreover, it can be shown that

$$PL = TDT^{-1} \tag{4.64}$$

where $D = diag(p_1, -p_2)$ with p_1, p_2 positive and T belongs to $U(1,1)$. By definition $T \in U(1,1)$ when

$$T^* LT = L \to T \in SU(1,1). \tag{4.65}$$

The proof is similar to that for Hermitian matrices. We shall regard PL as our new integration variable.

Note that (4.63) can be written in terms of p_1, p_2 except for $\epsilon\, tr\, P$ which will involve the integral over $SU(1,1)$. The p_1, p_2 are analogous to the radial variable r introduced in Sect. 4.6. Converting to the new coordinate system our measure becomes

$$(p_1 + p_2)^2 dp_1 dp_2 d\mu(T) \tag{4.66}$$

where $d\mu(T)$ the Haar measure on U(1,1). For large N, the p variables are approximately given by the complex saddle point

$$p_1 = -iE/2 + \rho(E), \quad -p_2 = -iE/2 - \rho(E), \qquad \rho(E) \equiv \sqrt{1-(E/2)^2}. \tag{4.67}$$

The p variables fluctuate only slightly about (4.67) while the T matrix ranges over the symmetric space $SU(1,1)/U(1)$ and produces a singularity for small ϵ. With the p_1, p_2 fixed the only remaining integral is over $SU(1,1)$. Thus from (4.64) we have:

$$Q \equiv PL \approx \rho(E)\; TLT^{-1} + iE/2.$$

If we set $\epsilon = \bar{\epsilon}/N$ and take the limit of large N we see that (4.63) is given by

$$\int e^{-\bar{\epsilon}\rho(E)\, tr(LS)}\, d\mu(T), \quad where \quad S \equiv TLT^{-1}. \tag{4.68}$$

This is the basis for the sigma model. The second term iE above is independent of T so it is dropped in (4.68).

4.7.1 Random Band Case

The band version or Wegner's N orbital [41] version of such hyperbolic sigma models was studied in [47]. The physical justification of this sigma model and also the Efetov SUSY sigma model comes from a granular picture of matter. The idea is that a metal consists of grains (of size $N \gg 1$). Within the grains there is mean field type interaction and there is a weak coupling across the neighboring grains. As in (4.68) if the grain interactions are scaled properly and $N \to \infty$ we will obtain a sigma model.

For each lattice site $j \in \Lambda \subset Z^d$ we define a new spin variable given by

$$S_j = T_j^{-1} L T_j \quad and \quad P_j L \approx \rho(E)\, S_j .$$

Note that $S_j^2 = 1$ and S_j naturally belongs to $SU(1,1)/U(1)$. This symmetric space is isomorphic to the hyperbolic upper half plane. In the last equation we have used the form of the p_i given as above. The imaginary part of the p_1 and $-p_2$ are equal at the saddle so that T and T^{-1} cancel producing only a trivial contribution. The explicit dependence on E only appears through $\rho(E)$.

There is a similar picture for the square of the average determinant using Grassmann integration. We can integrate out the Grassmann fields and in the sigma model approximation obtain an integral over the symmetric space $SU(2)/U(1) = S^2$—this gives the classical Heisenberg model.

The action of the hyperbolic sigma model, see [47], on the lattice is

$$A(S) = \beta \sum_{j \sim j'} tr\, S_j S_{j'} + \epsilon \sum_j tr\, L S_j . \tag{4.69}$$

The notation $j \sim j'$ denotes nearest neighbor vertices on the lattice. The Gibbs weight is proportional to $e^{-A(S)} d\mu(T)$. The integration over $SU(1,1)$ is divergent unless $\epsilon > 0$. The last term above is a symmetry breaking term analogous to a magnetic field . For RBM, $\beta \approx W^2 \rho(E)^2$.

To parametrize the integral over SU(1,1) we use *horospherical* coordinates $(s_j, t_j) \in \mathbb{R}^2$ given by (4.9). In this coordinate system, the action takes the form:

$$A(s,t) = \beta \sum_{j \sim j'} [cosh(t_j - t_{j'}) + \frac{1}{2}(s_j - s_{j'})^2 e^{(t_j + t_{j'})}] + \epsilon \sum_{j \in \Lambda} [cosh(t_j) + \frac{1}{2} s^2 e^{t_j}]. \tag{4.70}$$

Equivalently, if h_j are the hyperbolic spins appearing in (4.8) note that we have

$$h_j \cdot h_{j'} = z_j z_{j'} - x_j x_{j'} - y_j y_{j'} = cosh(t_j - t_{j'}) + \frac{1}{2}(s_j - s_{j'})^2 e^{(t_j + t_{j'})}.$$

The symmetry breaking term is just $\epsilon \sum z_j$. There is a symmetry $t_j \to t_j + \gamma$ and $s_j \to s_j\, e^{-\gamma}$ which leaves the action invariant when $\epsilon = 0$. Note that $A(s,t)$

is quadratic in s_j. Let us define the quadratic form associated to the s variables in (4.70):

$$[v\,;D_{\beta,\varepsilon}(t)\,v]_\Lambda = \beta \sum\nolimits_{(i\sim j)} \mathrm{e}^{t_i+t_j}(v_i - v_j)^2 + \varepsilon \sum\nolimits_{k\in\Lambda} \mathrm{e}^{t_k} v_k^2 \tag{4.71}$$

where v is a real vector, v_j, $j \in \Lambda$. Integration over the s fields produces $\det^{-1/2}(D_{\beta,\varepsilon}(t))$. The determinant is positive but non local in t. Thus the total *effective action* is given by

$$A(t) = \sum_{j\sim j'} \{\beta cosh(t_j - t_{j'})\} + \frac{1}{2}\ln\det(D_{\beta,\epsilon}(t)) + \epsilon \sum_j cosh(t_j). \tag{4.72}$$

The quadratic form $D_{\beta,\varepsilon}(t)$ will also appear in the SUSY sigma model defined later. Note that if $t = 0$, $D_{\beta,\varepsilon}(t) = -\beta\Delta + \epsilon$. For $t \neq 0$, $D_{\beta,\varepsilon}(t)$ is a finite difference elliptic operator which is the generator of a *random walk in a random environment.* The factor $e^{(t_j+t_{j'})}$ is the conductance across the edge (j, j'). Note that since t_j are not bounded, $D_{\beta,\varepsilon}(t)$ is *not* uniformly elliptic.

Lemma. $Det(D_{\beta,\varepsilon}(t))$ is a log convex function of the t variables.

Following D. Brydges, this a consequence of the matrix tree theorem which expresses the determinant of as a sum:

$$\sum_{\mathcal{F},\mathcal{R}} \prod_{j,j'\in\mathcal{F}} \beta\, e^{t_j+t_{j'}} \prod_{j\in\mathcal{R}} \epsilon\, e^{t_j}$$

where the sum above ranges over spanning rooted forests on Λ. See [1] for a proof of such identities.

Thus the effective action A (4.72) is convex and in fact its Hessian $A''(t) \geq -\beta\Delta + \epsilon$. The sigma model can now be analyzed using Brascamp–Lieb inequalities and a Ward identity. Its convexity will imply that this model does not have a phase transition in three dimensions.

Theorem (Brascamp–Lieb [5]). Let A(t) be a real convex function of t_j, $j \in \Lambda$ and v_j be a real vector. If the Hessian of the action A is convex $A''(t) \geq K > 0$ then

$$< e^{([v\,;t]-<[v\,;t]>)} >_A \leq e^{\frac{1}{2}[v\,;K^{-1}v]}. \tag{4.73}$$

Here K is a positive matrix independent of t and $<>_A$ denotes the expectation with respect to the normalized measure $Z_\Lambda^{-1}e^{-A(t)}D\,t$. Note if A is quadratic in t, (4.73) is an identity.

Note that in 3D the Brascamp–Lieb inequality only gives us estimates on functions of $([v\,;t] - <[v\,;t]>)$. Below we obtain estimates on $<[v\,;t]>$ by using Ward identities and translation invariance. Then the above theorem will yield a bound on moments of the local conductance $e^{(t_j+t_{j'})}$. These bounds on the conductance imply that in three dimensions we have "diffusion" with respect to the measure defined via (4.70). See the discussion following (4.87).

Theorem **([47]).** In the 3D hyperbolic sigma model described by (4.72), all moments of the form $< cosh^p(t_0) >$ are bounded for all β. The estimates are uniform in ϵ provided we first take the limit $\Lambda \to \mathbb{Z}^3$.

Proof. First note that if we let $v_j = p\delta_0(j)$ then in three dimensions since $K = -\beta\Delta + \epsilon$ we have $[v\,;\,K^{-1}v] \leq Const$ uniformly as $\epsilon \to 0$. To estimate the average $< t_0 >$ we apply the Ward identity:

$$2 < sinh(t_0) >=< s_0^2 e^{t_0} >=< D_{\beta,\varepsilon}^{-1}(0,0)e^{t_0} > . \tag{4.74}$$

To obtain this identity we apply $t_j \to t_j + \gamma$ and $s_j \to s_j\, e^{-\gamma}$ to (4.70) then differentiate in γ and set $\gamma = 0$. We have also assumed translation invariance. Since the right side of (4.74) is positive, if we multiply the above identity by $e^{-<t_0>}$ we have

$$< e^{(t_0 - <t_0>)} >=< e^{-t_0 - <t_0>} > + e^{-<t_0>} < s_0^2 e^{t_0} > \geq e^{-2<t_0>}$$

where the last bound follows from Jensen's inequality. The left side is bounded by Brascamp–Lieb by taking $v_j = \delta_0(j)$. Thus we get an upper bound on $- < t_0 >$ and on $< e^{-t_0} >$. To get the upper bound on $< t_0 >$ we use the inequality

$$e^{t_0} D_{\beta,\varepsilon}^{-1}(0,0) \leq e^{t_0}\, \beta \sum_{j \sim j'} e^{-t_j - t_{j'}}\, (G_0(0,j) - G_0(0,j'))^2 + O(\epsilon)$$

where $G_0 = (-\beta\Delta + \epsilon)^{-1}$. See (4.121) in Appendix 4.12 for a precise statement and proof of this inequality. The sum is convergent in 3D since the gradient of $G_0(0,j)$ decays like $|j|^{-2}$. Multiplying (4.74) by $e^{<t_0>}$ we get

$$< e^{(t_0 + <t_0>)} > \leq \beta \sum_{j \sim j'} < e^{-t_j - t_{j'}}\, e^{(t_0 + <t_0>)} > (G_0(0,j) - G_0(0,j'))^2$$
$$+ < e^{-(t_0 - <t_0>)} > + O(\epsilon).$$

By translation invariance $< t_j >=< t_0 >$, the exponent on the right side has the form $([v\,;\,t] - < [v\,;\,t] >)$ and thus the Brascamp–Lieb inequality gives us a bound on the right side and a bound on $< t_0 >$ by Jensen's inequality. Above we have ignored the $O(\epsilon)$ term. If we include it then we obtain an inequality of the form $e^{2<t_0>} \leq C + \epsilon e^{<t_0>}$ and the bound for $e^{<t_0>}$ still follows. Since we now have estimates on $< |t_j| >$, the desired estimate on $< cosh^p(t_j) >$ follows from the Brascamp–Lieb inequality.

Remark. The effective action for the SUSY hyperbolic sigma model, $H^{2|2}$, described in Sect. 4.8, looks very much like the formula above except that the coefficient of ln det is replaced by $-1/2$. For this case the action is *not convex* and so phase transitions are not excluded. In fact in 3D for small β there is localization and $< e^{-t_0} >$ will diverge as $\epsilon \to 0$. Note that the above results and those described in Sect. 4.8 rely heavily on the use of horospherical coordinates.

4.8 Phase Transition for a SUSY Hyperbolic Sigma Model

In this section we study a simpler version of the Efetov sigma models introduced by Zirnbauer [52]. This model is the $\mathbb{H}^{2|2}$ model mentioned in the introduction. This model is expected to qualitatively reflect the phenomenology of Anderson's tight binding model. The great advantage of this model is that the Grassmann degrees of freedom can be explicitly integrated out to produce a real effective action in bosonic variables. Thus probabilistic methods can be applied. In 3D we shall sketch the proof [12, 14] that this model has the analog of the *Anderson transition*. The analysis of the phase transition relies heavily on Ward identities and on the study of a random walk in a strongly correlated random environment.

In order to define the $\mathbb{H}^{2|2}$ sigma model, let u_j be a vector at each lattice point $j \in \Lambda \subset \mathbb{Z}^d$ with three bosonic components and two fermionic components

$$u_j = \left(z_j, x_j, y_j, \xi_j, \eta_j\right) ,$$

where ξ, η are odd elements and z, x, y are even elements of a real Grassmann algebra. The scalar product is defined by

$$(u, u') = -zz' + xx' + yy' + \xi\eta' - \eta\xi' , \qquad (u, u) = -z^2 + x^2 + y^2 + 2\xi\eta$$

and the action is obtained by summing over nearest neighbors $j, j\,'$

$$\mathscr{A}[u] = \frac{1}{2} \sum_{(j,j\,')\in\Lambda} \beta(u_j - u_{j\,'}, u_j - u_{j\,'}) + \sum_{j\in\Lambda} \varepsilon_j(z_j - 1) . \tag{4.75}$$

The sigma model constraint, $(u_j, u_j) = -1$, is imposed so that the field lies on a SUSY hyperboloid, $\mathbb{H}^{2|2}$.

We choose the branch of the hyperboloid so that $z_j \geq 1$ for each j. It is very useful to parametrize this manifold in horospherical coordinates:

$$x = \sinh t - e^t \left(\tfrac{1}{2}s^2 + \bar{\psi}\psi\right) , \quad y = se^t, \quad \xi = \bar{\psi}e^t, \quad \eta = \psi e^t,$$

and

$$z = \cosh t + e^t \left(\tfrac{1}{2}s^2 + \bar{\psi}\psi\right)$$

where t and s are even elements and $\bar{\psi}$, ψ are odd elements of a real Grassmann algebra.

In these coordinates, the sigma model action is given by

$$\mathscr{A}[t,s,\psi,\bar\psi] = \sum_{(ij)\in\Lambda} \beta(\cosh(t_i - t_j) - 1)$$
$$+\tfrac{1}{2}[s;D_{\beta,\varepsilon}s] + [\bar\psi;D_{\beta,\varepsilon}\psi] + \sum_{j\in\Lambda} \varepsilon_j(\cosh t_j - 1). \quad (4.76)$$

We define the corresponding expectation by $< \cdot > = < \cdot >_{\Lambda,\beta,\epsilon}$.

Note that the action is quadratic in the Grassmann and s variables. Here $D_{\beta,\varepsilon} = D_{\beta,\varepsilon}(t)$ is the generator of a *random walk in random environment*, given by the quadratic form

$$[v;D_{\beta,\varepsilon}(t)\,v]_\Lambda \equiv \beta\sum\nolimits_{(j\sim j')} e^{t_j+t_{j'}}(v_j - v_{j'})^2 + \sum\nolimits_{k\in\Lambda} \varepsilon_k\, e^{t_k} v_k^2 \quad (4.77)$$

as it is in (4.71). The weights, $e^{t_j+t_{j'}}$, are the *local conductances* across a nearest neighbor edge j, j'. The $\varepsilon_j\, e^{t_j}$ term is a killing rate for the walk at j. For the random walk starting at 0 without killing, we take $\epsilon_0 = 1$ and $\epsilon_j = 0$ otherwise. For the random band matrices if we set $\epsilon_j = \epsilon$ then ϵ may be thought of as the imaginary part of the energy.

After integrating over the Grassmann variables ψ, $\bar\psi$ and the variables $s_j \in \mathbb{R}$ in (4.76) we get the effective bosonic field theory with action $\mathscr{E}_{\beta,\varepsilon}(t)$ and partition function

$$Z_\Lambda(\beta,\epsilon) = \int e^{-\mathscr{E}_{\beta,\epsilon}(t)} \prod e^{-t_j}\,dt_j \equiv \int e^{-\beta\mathscr{L}(t)} \cdot [\det D_{\beta,\varepsilon}(t)]^{1/2} \prod_j e^{-t_j}\frac{dt_j}{\sqrt{2\pi}}. \quad (4.78)$$

where

$$\mathscr{L}(t) = \sum_{j\sim j'} [cosh(t_j - t_{j'}) - 1] + \sum_j \frac{\varepsilon_j}{\beta}[(\cosh(t_j) - 1]. \quad (4.79)$$

Note that the determinant is a positive but nonlocal functional of the t_j hence the effective action, $\mathscr{E} = \mathscr{L} - 1/2 \ln DetD_{\beta,\epsilon}$, is also nonlocal. The additional factor of e^{-t_j} in (4.78) arises from a Jacobian. Because of the internal supersymmetry, we know that for all values of β, ε the partition function

$$Z(\beta,\varepsilon) \equiv 1. \quad (4.80)$$

This identity holds even if β is edge dependent.

The analog of the Green's function $< |G(E_\epsilon;0,x)|^2 >$ of the Anderson model is the average of the Green's function of $D_{\beta,\varepsilon}$,

$$< s_0 e^{t_0} s_x e^{t_x} > (\beta,\varepsilon) = < e^{(t_0+t_x)} D_{\beta,\varepsilon}(t)^{-1}(0,x) > (\beta,\varepsilon) \equiv \mathscr{G}_{\beta,\varepsilon}(0,x) \quad (4.81)$$

where the expectation is with respect to the SUSY statistical mechanics weight defined above. The parameter $\beta = \beta(E)$ is roughly the bare conductance across an edge and we shall usually set $\varepsilon = \varepsilon_j$ for all j. In addition to the identity $Z(\beta, \varepsilon) \equiv 1$, there are additional Ward identities

$$< e^{t_j} >\equiv 1, \qquad \varepsilon \sum_x \mathcal{G}_{\beta,\varepsilon}(0,x) = 1 \tag{4.82}$$

which hold for all values of β and ε. The second identity above corresponds to the Ward identity (4.4).

Note that if $|t_j| \le Const$, then the conductances $e^{t_j+t_{j'}}$ are uniformly bounded from above and below and

$$D_{\beta,\varepsilon}(t)^{-1}(0,x) \approx (-\beta\Delta + \varepsilon)^{-1}(0,x)$$

is the diffusion propagator. Thus localization can only occur due to large deviations of the t field.

An alternative Schrödinger like representation of (4.81) is given by

$$\mathcal{G}_{\beta,\varepsilon}(0,x) =< \tilde{D}^{-1}_{\beta,\varepsilon}(t)(0,x) > \tag{4.83}$$

where

$$e^{-t} D_{\beta,\varepsilon}(t) e^{-t} \equiv \tilde{D}_{\beta,\varepsilon}(t) = -\beta\Delta + \beta V(t) + \varepsilon\, \mathrm{e}^{-t}, \tag{4.84}$$

and $V(t)$ is a diagonal matrix (or "potential") given by

$$V_{jj}(t) = \sum_{i:|i-j|=1} (\mathrm{e}^{t_i - t_j} - 1).$$

In this representation, the potential is highly correlated and $\tilde{D} \ge 0$ as a quadratic form.

Some insight into the transition for the $\mathbb{H}^{2|2}$ model can be obtained by finding the configuration $t_j = t^*$ which minimizes the effective action $\mathcal{E}_{\beta,\varepsilon}(t)$ appearing in (4.78). It is shown in [14] that this configuration is unique and does not depend on j. For large β, it is given by

$$\text{1D:} \quad \varepsilon\, \mathrm{e}^{-t^*} \simeq \beta^{-1}, \qquad \text{2D:} \quad \varepsilon\, \mathrm{e}^{-t^*} \simeq \mathrm{e}^{-\beta}, \qquad \text{3D:} \qquad t^* \simeq 0. \tag{4.85}$$

Note that in one and two dimensions, t^* depends sensitively on ϵ and that negative values of t_j are favored as $\varepsilon \to 0$. This means that at t^* a mass εe^{-t^*} in (4.84) appears even as $\varepsilon \to 0$. Another interpretation is that the classical conductance $e^{t_j+t_{j'}}$ should be small in *some sense*. This is a somewhat subtle point. Due to large deviations of the t field in 1D and 2D, $<e^{t_j+t_{j'}}>$ is expected to diverge, whereas $< e^{t_j/2} >$ should become small as $\varepsilon \to 0$. One way to adapt the saddle approximation so that it is sensitive to different observables is to include the observable when computing the saddle point. For example, when taking the expectation of e^{pt_0}, the saddle is only slightly changed when $p = 1/2$ but for $p = 2$ it will give a divergent contribution when there is localization.

When β is small, $\varepsilon e^{-t^*} \simeq 1$ in any dimension. Thus the saddle point t^* suggests localization occurs in both 1D and 2D for all β and in 3D for small β. In 2D, this agrees with the predictions of localization by Abrahams, Anderson, Licciardello and Ramakrishnan [2] at any nonzero disorder. Although the saddle point analysis has some appeal, it is not easy to estimate large deviations away from t^* in one and two dimensions. In 3D, large deviations away from $t^* = 0$ are controlled for large β. See the discussion below.

The main theorem established in [14] states that in 3D, fluctuations around $t^* = 0$ are rare. Let $G_0 = (-\beta\Delta + \epsilon)^{-1}$ be the Green's function for the Laplacian.

Theorem 14 *If $d \geq 3$, and the volume $\Lambda \to \mathbb{Z}^d$, there is a $\bar{\beta} \geq 0$ such that for $\beta \geq \bar{\beta}$ then for all j,*

$$< cosh^8(t_j) > \leq Const \tag{4.86}$$

where the constant is uniform in ϵ. This implies diffusion in a quadratic form sense: Let $\mathcal{G}$ be given by (4.81) *or* (4.83). *There is a constant C so that we have the quadratic form bound*

$$\frac{1}{C}[f; G_0 f] \leq \sum_{x,y} \mathcal{G}_{\beta,\varepsilon}(x,y)\, f(x) f(y) \leq C[f; G_0 f], \tag{4.87}$$

where f(x) is nonnegative function.

Remarks. A weaker version of the lower bound in (4.87) appears in [14]. Yves Capdeboscq (private communication) showed (4.87) follows directly from the (4.86). This proof is explained in Appendix 4.12. The power 8 can be increased by making β larger. One expects pointwise diffusive bounds on $\mathcal{G}_{\beta,\varepsilon}(x,y)$ to hold. However, in order to prove this one needs to show that the set $(j : |t_j| \leq M)$ percolates in a strong sense for some large M. This is expected to be true but has not yet been mathematically established partly because of the strong correlations in the t field.

The next theorem establishes localization for small β in any dimension. See [12].

Theorem 12 *Let $\varepsilon_x > 0$, $\varepsilon_y > 0$ and $\sum_{j\in\Lambda} \varepsilon_j \leq 1$. Then for all $0 < \beta < \beta_c$ (β_c defined below) the correlation function $\mathcal{G}_{\beta,\varepsilon}(x,y)$,* (4.83), *decays exponentially with the distance $|x-y|$. More precisely:*

$$\mathcal{G}_{\beta,\varepsilon}(x,y) = \langle \tilde{D}^{-1}_{\beta,\varepsilon}(t)(x,y)\rangle \leq C_0 \left(\varepsilon_x^{-1} + \varepsilon_y^{-1}\right) \left[I_\beta\, e^{\beta(c_d-1)}\, c_d\right]^{|x-y|} \tag{4.88}$$

where $c_d = 2d - 1$, C_0 *is a constant and*

$$I_\beta = \sqrt{\beta} \int_{-\infty}^{\infty} \frac{dt}{\sqrt{2\pi}} e^{-\beta(\cosh t - 1)} .$$

Finally β_c *is defined so that:*

$$\left[I_\beta \, e^{\beta(c_d-1)} \, c_d\right] < \left[I_{\beta_c} e^{\beta_c(c_d-1)} c_d\right] = 1 \quad \forall \beta < \beta_c.$$

These estimates hold uniformly in the volume. Compare (4.88) *with* (4.2).

Remarks. The first proof of localization for the $\mathbb{H}^{2|2}$ model in 1D was given by Zirnbauer in [52]. Note that in 1D, $c_d - 1 = 0$ and exponential decay holds for all $\beta < \infty$ and the localization length is bounded by β when β is large. One expects that for 1D strips of width $|S|$ and β large, the localization length is proportional to $\beta|S|$. However, this has not yet been proved. The divergence in ε^{-1} is compatible with the Ward identity (4.4) and is a signal of localization.

4.8.1 Role of Ward Identities in the Proof

The proof of Theorems 3 and 4 by rely heavily on Ward identities arising from internal SUSY. These are described below. For Theorem 14 we use Ward identities to bound fluctuations of the t field by getting bounds in 3D on $< cosh^m(t_i - t_j) >$. This is done by induction on the distance $|i - j|$. Once these bounds are established we use ϵ to get bounds for $< cosh^p \, t >$. For Theorem 12 we use the fact that for any region Λ, the partition function $Z_\Lambda = 1$.

If an integrable function S of the variables $x, y, z, \psi, \bar{\psi}$ is supersymmetric, i.e., it is invariant under transformations preserving

$$x_i x_j + y_i y_j + \bar{\psi}_i \psi_j - \psi_i \bar{\psi}_j$$

then $\int S = S(0)$. In horospherical coordinates the function S_{ij} given by

$$S_{ij} = B_{ij} + e^{t_i + t_j}(\bar{\psi}_i - \bar{\psi}_j)(\psi_i - \psi_j) \tag{4.89}$$

where

$$B_{ij} = cosh(t_i - t_j) + \frac{1}{2} e^{t_i + t_j}(s_i - s_j)^2 \tag{4.90}$$

is supersymmetric. If i and j are nearest neighbors, $S_{ij} - 1$ is a term in the action $\mathscr{A}$ given in (4.76) and it follows that the partition function $Z_\Lambda(\beta, \epsilon) \equiv 1$. More generally for each m we have

$$< S_{ij}^m >_{\beta,\epsilon} = < B_{ij}^m \left[1 - m B_{ij}^{-1} e^{t_i + t_j}(\bar{\psi}_i - \bar{\psi}_j)(\psi_i - \psi_j)\right] >_{\beta,\epsilon} \equiv 1. \tag{4.91}$$

Here we have used that $S_{ij}^m e^{-\mathscr{A}_\epsilon}$ is integrable for $\epsilon > 0$. The integration over the Grassmann variables in (4.91) is explicitly given by

$$G_{ij} = \frac{e^{t_i+t_j}}{B_{ij}} \left[(\delta_i - \delta_j);\ D_{\beta,\varepsilon}(t)^{-1}(\delta_i - \delta_j)\right]_\Lambda \tag{4.92}$$

since the action is quadratic in $\bar{\psi}, \psi$. Thus we have the identity

$$< B_{ij}^m(1 - mG_{ij}) > \equiv 1. \tag{4.93}$$

Note that $0 \le cosh^m(t_i - t_j) \le B_{ij}^m$. From the definition of $D_{\beta,\varepsilon}$ given in (4.77) we see that for large β, G in (4.92) is typically proportional to $1/\beta$ in 3D. However, there are rare configurations of $t_k \ll -1$ with k on a surface separating i and j for which G_{ij} can diverge as $\epsilon \to 0$. In 2D, G_{ij} grows logarithmically in $|i-j|$ as $\epsilon \to 0$ even in the ideal case $t \equiv 0$.

Our induction starts with the fact that if i, j are nearest neighbors then it is easy to show that G_{ij} is less than β^{-1} for all t configurations. This is because of the factor $e^{t_i+t_j}$ in (4.92).

If $|i - j| > 1$ and $m\ G_{ij} \le 1/2$, then (4.93) implies that

$$0 \le\ < cosh^m(t_i - t_j) > \le < B_{ij}^m > \le 2.$$

It is not difficult to show that one can get bounds on G_{ij} depending only on the t fields in a double cone with vertices at i and j. In fact for k far way from i, j, the dependence of G_{ij} on t_k is mild. Nevertheless, there is still no uniform bound on G_{ij} due of t fluctuations. We must use induction on $|i - j|$ and SUSY characteristic functions of the form $\chi\{S_{ij} \le r\}$, to prove that configurations with $1/2 \le m\ G_{ij}$, with $\beta \gg m \gg 1$, are rare for large β in 3D. We combine these estimates with the elementary inequality $B_{ik}B_{kj} \ge 2B_{ij}$ for $i, j, k \in \mathbb{Z}^d$ to go to the next scale. See [14] for details.

The proof of the localized phase is technically simpler than the proof of Theorem 14. Nevertheless, it is of some interest because it shows that $\mathbb{H}^{2|2}$ sigma model reflects the localized as well as the extended states phase in 3D. The main idea relies on the following lemma. Let M be an invertible matrix indexed by sites of Λ and let γ denote a self avoiding path starting at i and ending at j. Let M_{ij}^{-1} be matrix elements of the inverse and let M_{γ^c} be the matrix obtained from M by striking out all rows and columns indexed by the vertices covered by γ.

Lemma. Let M and M_{γ^c} be as above, then

$$M_{ij}^{-1}\det M = \sum_{\gamma\ i,j} \left[(-M_{ij_1})(-M_{j_1j_2})\cdots(-M_{j_mj})\right] \det M_{\gamma^c}$$

where the sum ranges over all self-avoiding paths γ connecting i and j, $\gamma_{i,j} = (i, j_1, \ldots j_m, j)$, with $m \ge 0$.

Apply this lemma to

$$M = e^{-t} D_{\beta,\varepsilon}(t) e^{-t} \equiv \tilde{D}_{\beta,\varepsilon}(t) = -\beta\Delta + V(t) + \varepsilon\, \mathrm{e}^{-t} \tag{4.94}$$

and notice that for this M, for all non-zero contributions, γ are nearest neighbor self-avoiding paths and that each step contributes a factor of β. To prove the lemma, write the determinant of M as a sum over permutations of $j \in \Lambda$ indexing M. Each permutation can be expressed as a product of disjoint cycles covering Λ. Let E_{ij} denote the elementary matrix which is 0 except at ij place where it is 1. The derivative in s of $\det(M + sE_{ji})$ at $s = 0$ equals $M_{ij}^{-1}\det M$ and selects the self-avoiding path in the cycle containing j and i. The other loops contribute to $\det M_{\gamma^c}$. By (4.83) and (4.94) we have

$$\mathcal{G}_{\beta,\varepsilon}(x,y) = < M_{xy}^{-1} > = \int e^{-\beta\mathcal{L}(t)} M_{xy}^{-1} [\det M]^{1/2} \prod_j \frac{dt_j}{\sqrt{2\pi}}. \tag{4.95}$$

Note the factors of e^{-t_j} appearing in (4.78) have been absorbed into the determinant. Now write

$$M_{xy}^{-1}[\det M]^{1/2} = \sqrt{M_{xy}^{-1}}\sqrt{M_{xy}^{-1}\det M}.$$

The first factor on the right hand side is bounded by $\epsilon_x^{-1/2}\, e^{t_x/2} + \epsilon_y{}^{-1/2} e^{t_y/2}$. For the second factor, we use the lemma. Let $\mathcal{L} = \mathcal{L}_\gamma + \mathcal{L}_{\gamma^c} + \mathcal{L}_{\gamma,\gamma^c}$ where $\mathcal{L}_\gamma$ denotes the restriction of $\mathcal{L}$ to γ. Then by supersymmetry,

$$\int e^{-\beta\mathcal{L}_{\gamma^c}} [\det M_{\gamma^c}]^{1/2} \prod_j \frac{dt_j}{\sqrt{2\pi}} \equiv 1$$

we can bound

$$0 \le \mathcal{G}_{\beta,\varepsilon}(x,y) \le \sum_{\gamma_{xy}} \sqrt{\beta}^{|\gamma_{xy}|} \int e^{-\beta\mathcal{L}_\gamma + \beta\mathcal{L}_{\gamma,\gamma^c}} [\epsilon_x^{-1/2}\, e^{t_x/2} + \epsilon_y{}^{-1/2} e^{t_y/2}] \prod_j \frac{dt_j}{\sqrt{2\pi}}$$

where $|\gamma_{xy}|$ is the length of the self-avoiding path from x to y. The proof of Theorem 12 follows from the fact that the integral along γ is one dimensional and can be estimated as a product. See [12] for further details.

4.9 Efetov's Sigma Model

In this final section we present a very brief description of the Efetov sigma model for Hermitian matrices, following [17, 18, 36, 51]. This sigma model is the basis for most random band matrix calculations in theoretical physics. Unfortunately, the mathematical analysis of this model is still limited to 1D. Even in this case our analysis is far from complete. The ideas of this section are quite similar to those in Sect. 4.7 leading to (4.68) and (4.69) except that we now include Grassmann variables.

In order to obtain information about averages of the form (4.1) we introduce a field with four components

$$\Phi_j = (z_j, w_j, \psi_j, \chi_j) \tag{4.96}$$

where z, w are complex fields and ψ, χ are Grassmann fields. Let $L = diag(1, -1, 1, 1)$, and $\Lambda = diag(1, -1, 1, -1)$. For a Hermitian matrix H, define the action

$$A(E, \epsilon) = \Phi^* \cdot L \{i(H - E) + \epsilon \Lambda\} \Phi. \tag{4.97}$$

Note that the signature of L is chosen so that the z and w variables appear as complex conjugates of each other as they do in (7.1). Then we have the identity:

$$|(E - i\epsilon - H)^{-1}(0, j)|^2 = \int z_0 \bar{z}_j w_0 \bar{w}_j \, e^{-A(E,\epsilon)} \, D\Phi \tag{4.98}$$

where

$$D\Phi \equiv Dz \, Dw \, D\psi \, D\chi.$$

Without the observable $z_0 \bar{z}_j w_0 \bar{w}_j$, $\int e^{-A} = 1$. The integral over Gaussian H can be calculated as in (4.55) and will produce a quartic interaction in Φ. However, now the Hubbard–Stratonovich transformation, which is usually used in the Bosonic sector, involves a subtle analytic continuation first worked out in [41], see also [28, 51].

Now let us define a matrix of the form

$$M = \begin{pmatrix} [BB] & [BF] \\ [FB] & [FF] \end{pmatrix} \tag{4.99}$$

where each block is a 2×2 matrix. M will be called a *supermatrix* if the diagonal blocks, BB and FF are made of commuting (even elements) variables while the off diagonal blocks FB and BF consist of odd elements in the Grassmann algebra. Define the supertrace

$$Str(M) \equiv Tr([BB] - [FF]).$$

Note that for supermatrices A and B we have $Str(AB) = Str(BA)$. We define the adjoint of a supermatrix M by

$$M^\dagger = \begin{pmatrix} [BB]^* & [FB]^* \\ -[BF]^* & [FF]^* \end{pmatrix}.$$

The symbol * denotes the usual transpose followed by conjugation. For Grassmann variables we have $\overline{\psi_a \psi_b} = \bar{\psi}_a \bar{\psi}_b$. But $\bar{\bar{\psi}} = -\psi$ so that † is an involution and

$$\Phi_1^\dagger(M\Phi_2) = (M^\dagger\Phi_1)^\dagger\Phi_2.$$

For $\epsilon = 0$ the action is invariant under the action of matrices T in $SU(1;1|2)$ which satisfy:

$$T^\dagger LT = L\,. \tag{4.100}$$

As in (4.68) and (4.69) the spin or field of the sigma model is given by matrices

$$S_j = T_j^{-1}\Lambda T_j \tag{4.101}$$

as T ranges over $SU(1;1|2)$. It is the orbit of a critical point, which is proportional to Λ under the action of $SU(1,1|2)$. Thus the matrix S ranges over a supersymmetric space $U(1,1|2)/(U(1|1)\times U(1,1))$. A general discussion of more general supersymmetric spaces appears in [53]. The SUSY sigma model has a Gibbs density defined by

$$\exp\{-\beta Str\sum_{j\sim j'}(S_j - S_{j'})^2 - \epsilon\, Str\sum_j \Lambda S_j\}. \tag{4.102}$$

In a one dimensional chain of length L with $\epsilon = 0$ except at the end points, the Grassmann variables can be explicitly integrated over producing the formula (4.8). In higher dimensions the action (4.102) is harder to analyze because of the Fermion–Boson coupling of the BB sector (hyperbolic sigma model) and the FF sector (Heisenberg model). An explicit parametrization of the 4×4 matrices S_j and integration measure is given in [17, 18, 36]. Fluctuations about the saddle should produce massless modes—Goldsone Bosons in 3D.

4.10 Appendix A: Gaussian and Grassmann Integration

Let $z = x + iy$ with $x, y \in \mathbb{R}$. Let dz = $dxdy/\pi$ and suppose $Re\, a > 0,\ a \in \mathbb{C}$. Then

$$\int e^{-a\bar{z}z}dz = \pi^{-1}\iint e^{-ar^2}rdrd\theta = a^{-1}. \tag{4.103}$$

Also

$$\frac{1}{\sqrt{2\pi}}\int e^{-ax^2/2}dx = a^{-1/2}.$$

In the multi-dimensional case let $z = (z_1, z_2, \ldots z_n)$, $z^* = \bar{z}^t$. For an $n\times n$ matrix A with $Re\, A > 0$ as a quadratic form

$$\int e^{-[z;\,Az]}Dz = (\det A)^{-1} \quad where \quad Dz = \prod_1^n dx_i dy_i/\pi. \tag{4.104}$$

Recall the notation $[z; Az] \equiv \sum \bar{z}_j A_{ij} z_j = z^* Az$. To prove the formula, first note that it holds in the Hermitian case by diagonalization. Since both sides of the equation are analytic in the matrix entries, and agree for the Hermitian case, the result follows by uniqueness of analytic continuation. The expectation with respect to A is defined by

$$< \cdot >_A \equiv \det(A) \int e^{-z^* Az} \ \cdot \ Dz$$

and from integration by parts, the pair correlation is given by

$$< z_j \bar{z}_k >_A = A^{-1}_{jk}. \tag{4.105}$$

Note that $< z_j z_k >_A = < \bar{z}_j \bar{z}_k >_A = 0$. This is because the integral is invariant under the global transform $z \to e^{i\phi} z$, $\bar{z} \to e^{-i\phi} \bar{z}$. The generating function is given by

$$< e^{z^* v + w^* z} >_A = e^{w^* A^{-1} v} = e^{[w; A^{-1} v]}.$$

This identity follows by changing variables: $z \to z - A^{-1} v$ and $\bar{z} \to \bar{z} - (A^t)^{-1} \bar{w}$.

For real variables x = $(x_1, \dots x_n)$ if A is symmetric and positive

$$\int e^{-[x;Ax]/2} \, Dx = (detA)^{-1/2} \quad where \quad Dx = \prod_i^n dx_i / \sqrt{2\pi}. \tag{4.106}$$

Its generating function $< e^{[x;y]} >_A = e^{[y;A^{-1} y]/2}$, is obtained by completing the square.

There are similar formulas for integration over $N \times N$ matrices:

$$\int e^{-NTrH^2/2} e^{iTrMH} \, DH = e^{-TrM^2/2N} \int e^{-NTrH^2/2} DH. \tag{4.107}$$

For the case of band matrices the generating function is

$$< e^{iTrHM} >= e^{-<(trHM)^2)/2>} = e^{-1/2 \sum J_{ij} M_{ji} M_{ij}}. \tag{4.108}$$

4.10.1 Grassmann Integration

Grassmann variables $\psi_i, \bar{\psi}_j$ are anti-commuting variables $1 \leq i, j \leq N$ satisfying $\psi_j^2 = \bar{\psi}_j^2 = 0, \, and \ \bar{\psi}_j \psi_i = -\psi_i \bar{\psi}_j$. Also

$$\psi_j \psi_i = -\psi_i \psi_j, \qquad \bar{\psi}_j \bar{\psi}_i = -\bar{\psi}_i \bar{\psi}_j. \tag{4.109}$$

The $\bar{\psi}_j$ is simply convenient notation for another independent family of Grassmann variables which anti-commute with ψ. Even monomials in the Grassmann variables and complex numbers commute with Grassmann variables. The polynomials in these variables form a Z_2 graded algebra, with the even and odd monomials belonging to the even and odd gradings respectively. One way to think about the

Grassmann variables is to let $\psi_j = dx_j$ and $\bar{\psi}_j = dy_j$ and consider the product as the wedge product in the theory of differential forms.

The Grassmann integral, defined below, plays an important role in many parts of physics. It is an extremely efficient and useful notation for the analysis of interacting Fermi systems, Ising models (Free Fermions), and SUSY. Although most of the time we shall eliminate the Grassmann variables by integrating them out, they are nevertheless an essential tool for obtaining the identities we shall analyze. See [1,4, 24,36,40] for more details about Grassmann integration.

We define integration with respect to

$$D\psi \equiv \prod_j^N d\bar{\psi}_j d\psi_j \tag{4.110}$$

as follows. For N = 1

$$\int (a\psi_1\bar{\psi}_1 + b\psi_1 + c\bar{\psi}_1 + d)\, D\psi = a.$$

The general rule is that the integral of a polynomial in 2N variables with respect to $D\psi$ equals the coefficient of the top monomial of degree 2N ordered as $\prod^N \psi_j\bar{\psi}_j$. Note that since the factors in the product are even, their order does not matter. Any element of the Grassmann algebra can be expressed as a polynomial and the top monomial can always be rearranged using the anti-commutation rules so that it coincides with $\prod^N \psi_j\bar{\psi}_j$.

To differentiate a Grassmann monomial, use the rule $\frac{\partial}{\partial\psi_j}\psi_k = \delta_{jk}$. The derivative anti-commutes with other Grassmann variables. We have

$$\frac{\partial}{\partial\psi_j}\psi_j \prod_{k\neq j}\psi_k = \prod_{k\neq j}\psi_k.$$

To differentiate a general monomial in ψ_j, use linearity and the anti-commutation relations so that it is of the above form. If ψ_j is not a factor then the derivative is 0.

For any $N \times N$ matrix A we have the following analog of Gaussian integration

$$\int e^{-[\psi;A\psi]}\, D\psi = detA \quad where \quad [\psi;A\psi] = \sum \bar{\psi}_i A_{ij}\psi_j. \tag{4.111}$$

Moreover,

$$< \psi_i\bar{\psi}_j > \equiv detA^{-1} \int \psi_i\bar{\psi}_j\, e^{-[\psi;A\psi]} D\psi = A_{ij}^{-1}. \tag{4.112}$$

More generally for a polynomial F in $\psi, \bar{\psi}$ we can integrate by parts to obtain

$$\int \psi_i \, F e^{-[\psi;A\psi]} D\psi = \sum_j A_{ij}^{-1} \int \{\frac{\partial}{\partial \bar{\psi}_j} F\} e^{-[\psi;A\psi]} D\psi. \tag{4.113}$$

To prove this formula we use

$$\sum_j A_{ij}^{-1} \int \frac{\partial}{\partial \bar{\psi}_j} (e^{-[\psi;A\psi]} F) D\psi = 0\,.$$

Let us establish (4.111) in the simplest case:

$$\int e^{-a\bar{\psi}_1\psi_1} D\psi = \int (1 - a\bar{\psi}_1\psi_1) D\psi = a \int \psi_1\bar{\psi}_1 D\psi = a.$$

Exercise. Show that if A is a 2×2 matrix, (4.111) holds.

To prove the general case note that the exponential can be written as a product $\prod_i (1 - \sum_j A_{ij}\bar{\psi}_i\psi_j)$ and we look at the terms:

$$\sum A_{1j_1} A_{2j_2} \dots A_{N,j_N} \int \bar{\psi}_1\psi_{j_1} \, \bar{\psi}_2\psi_{j_2} \dots \bar{\psi}_N\psi_{j_N} \; D\psi.$$

The j_i are distinct and hence are a permutation of $1 \dots N$. The integral then is the sign of the permutation and thus we obtain the determinant. The generating function is given by

$$< e^{\bar{\rho}^t\psi + \bar{\psi}^t\rho} >= detA^{-1} \int e^{-[\psi;A\psi]} \, e^{\bar{\rho}^t\psi + \bar{\psi}^t\rho} D\psi \; = e^{\bar{\rho}^t A^{-1}\rho} \tag{4.114}$$

where $\rho, \bar{\rho}$ are independent Grassmann variables.

4.10.2 Polar Coordinates for Grassmann and Bosonic Matrices

Grassmann integrals can also be expressed as integrals over unitary matrices. Let $d\mu(U)$ denote the Haar measure on $U(m)$. Given a family of Grassman variables $\bar{\psi}_j^\alpha, \psi_j^\alpha$ with $1 \le \alpha \le N \; and \; 1 \le j \le m$, define the matrix $M_{ij} = \psi_i \cdot \bar{\psi}_j$, where the dot product is the sum over α. Then for a smooth function F of M we have

$$\int F(M) \, D\psi = C_{N,m} \int F(U) \; \det(U)^{-N} \; d\mu(U) \tag{4.115}$$

where $U \in U(m)$. For example for $m = 1$

$$\int e^{a\psi\cdot\bar{\psi}} D\psi = a^N = \frac{N!}{2\pi} \int exp(ae^{i\theta}) e^{-iN\theta} d\theta.$$

It suffices to check (4.115) when $F = e^{trQ\,M}$ for a general matrix Q since any function can be approximated by linear combinations of such exponentials. From (4.111) the left side is equal to $\det(Q)^N$. These expressions agree since

$$\int e^{trQ\,U} \det(U)^{-N}\, d\mu(U) \propto \det(Q)^N.$$

To prove this statement first suppose that $Q = e^{itH}$ is a unitary matrix. Our identity now follows by the invariance of Haar measure. Since both sides are analytic in t and agree for t real, they agree for all matrices Q by taking t complex.

For bosonic fields the analog of the above relation is given by the generalized gamma function or Ingham–Siegel identity, see [29]:

$$\int_{P>0} e^{-trQ\,P} det(P)^{N-m}\, dP \propto det(Q)^{-N}.$$

Here P and Q denote positive Hermitian $m \times m$ matrices and dP denotes the flat measure restricted positive matrices. We must assume that $N \geq m$. Hence, if we set $M_{ij} = \bar{z}_i, \cdot z_j$ we have

$$\int F(M) Dz = C'_{N,m} \int_{P>0} F(P)\, \det(P)^{N-m}\, dP. \tag{4.116}$$

Note that $\det(P)^{-m} dP$ is proportional to the measure $d\mu(P)$ which is invariant under the transformation $P \to g^* P g$, for $g \in GL_m(\mathbb{C})$. For example dt/t is invariant under $t \to at$ for $a > 0$.

We may think of U and P as the polar or collective variable for the Grassmann and Bosonic variables respectively and their eigenvalues could be referred to as the radial coordinates.

Let $\Phi = (z, \psi)$ denote a column vector of with n bosonic components, z_j, and n Grassmann components, ψ_j. Define Φ^* to be the transpose conjugate of Φ. If M denotes a supermatrix with $n \times n$ blocks of the form (4.99), then the standard SUSY integration formula

$$\int e^{-\Phi^* M \Phi} Dz D\psi = SDet^{-1}(M)$$

holds, where the superdeterminant is given by

$$SDet(M) = det^{-1}([FF])\, det([BB] - [BF][FF]^{-1}[FB]).$$

This formula may be derived by first integrating out the ψ variables using (4.114) and then integrating the z variables. An equivalent formula given by

$$SDet(M) = det([BB])\, det^{-1}([FF] - [FB][BB]^{-1}[BF])$$

is obtained by reversing the order of integration. If A and B are supermatrices then $SDet(AB) = SDet(A)\, SDet(B)$ and ln $SDetA = StrA$.

Now consider a supermatrix with both Grassmann and bosonic components. For example, in the simplest case, (m = 1):

$$M = \begin{pmatrix} z\cdot\bar{z} & z\cdot\bar{\psi} \\ \psi\cdot\bar{z} & \psi\cdot\bar{\psi} \end{pmatrix}.$$

then for suitably regular F, the SUSY generalization of the above formulas is given by

$$\int F(M)\, DzD\psi \propto \int F(Q)\, SDet(Q)^N\, DQ.$$

Here Q is a 2 × 2 supermatrix and m = 1. For $m \geq 1$ of Q has the form

$$Q = \begin{pmatrix} P & \bar{\chi} \\ \chi & U \end{pmatrix}$$

with blocks of size $m \times m$ and

$$DQ \equiv dP\, d\mu(U)\, det^m(U)\, D\bar{\chi} D\chi$$

where $d\mu$ Haar measure on $U(m)$, and dP is the flat measure on positive Hermitian matrices. As a concrete example let us evaluate the integral using the above formula for m = 1 in the special case

$$\int e^{-a\bar{z}\cdot z + b\psi\bar{\psi}}\, Dz\, D\psi = b^N a^{-N} = \int e^{-ap+be^{i\theta}}\, SDet^N(Q)\, dp\, d\theta e^{i\theta} d\bar{\chi}\, d\chi$$

where

$$Sdet^N(Q) = p^N e^{-iN\theta}(1 - Np^{-1}\bar{\chi}e^{-i\theta}\chi).$$

Note that in the contribution to the integral, only the second term of $SDet$ above contributes. It is of top degree equal to $N\, p^{N-1}e^{-i(N+1)\theta}\chi\bar{\chi}$.

Remarks. The above integral identity is a simple example of *superbozonization*, See [33] for a more precise formulation and proof as well as references.

4.11 Appendix B: Formal Perturbation Theory

In this appendix we will explain some formal perturbation calculations for the average Green's function of the random Schrödinger matrix when the strength of the disorder λ is small. For a rigorous treatment of perturbation theory, see for example [22]. At present it is not known how to rigorously justify this perturbation theory when the imaginary part of the energy, ϵ is smaller than λ^4 . In fact the best rigorous estimates require $\epsilon \geq \lambda^p,\ p \leq 2+\delta,\ \delta > 0$. We shall explain some of the challenges in justifying the perturbation theory. Similar perturbative calculations can be done for random band matrices and in special cases these can be justified using SUSY as explained in Sect. 4.5.

Let us write the Green's function for the random Schrödinger matrix as

$$G(E_\epsilon; j,k) = [E_\epsilon + \Delta - \lambda^2\sigma(E_\epsilon) - \lambda v + \lambda^2\sigma(E_\epsilon)]^{-1}(j,k).$$

We assume that the potential v_j are independent Gaussian random variables of 0 mean and variance 1. Note that we have added and subtracted a constant $\lambda^2\sigma(E_\epsilon)$ and we shall now expand around

$$G_0(E_\epsilon; j,k) \equiv [E_\epsilon + \Delta - \lambda^2\sigma(E_\epsilon)]^{-1}.$$

The first order term in λ vanishes since v has mean 0. The basic idea is to choose the constant $\sigma(E_\epsilon)$ so that after averaging, the second order term in λ vanishes:

$$\lambda^2 < G_0 v G_0 v G_0 - G_0 \sigma G_0 >= 0 \quad hence \quad G_0(E_\epsilon; j,j) = \sigma(E_\epsilon) \tag{4.117}$$

holds for all j. This gives us an equation for $\sigma(E_\epsilon)$ which is the leading contribution to the self-energy. In field theory this is just self-consistent Wick ordering. When E lies in the spectrum of $-\Delta$, the imaginary part of $\sigma(E_\epsilon)$ does not vanish even as $\epsilon \to 0$. In fact, to leading order $Im\sigma(E)$ is proportional to the density of states for $-\Delta$. See (4.33). Thus to second order in perturbation theory we have $< G(E_\epsilon; j,k) >\approx G_0(E_\epsilon; j,k)$. Note that since the imaginary part of $\sigma(E_\epsilon) > 0$, $G_0(E_\epsilon; j,k)$ will decay exponentially fast in $\lambda^2|j-k|$ uniformly for $\epsilon > 0$.

Exponential decay is also believed to hold for $< G(E_\epsilon; j,k) >$ in all dimensions but has only been proved in one dimension. This decay is *not* related to localization and should also hold at energies where extended states are expected to occur. If λ is large and if v has a Gaussian distribution then it is easy to show that in any dimension the average Green's function decays exponentially fast. One way to see this is to make a shift in the contour of integration $v_j \to v_j + i\delta$. Since λ is large it will produce a large diagonal term. We get a convergent random walk expansion for $< G >$ by expanding in the off diagonal matrix elements.

Remark. Note that by the Ward identity (4.4) and (4.117) we have

$$\sum_j |G_0(E_\epsilon; 0,j)|^2(\lambda^2 Im\sigma(E_\epsilon) + \epsilon) = Im\sigma(E_\epsilon). \tag{4.118}$$

We now explain a problem in justifying perturbation theory. First note that one could proceed by defining higher order corrections to the self-energy, $\lambda^4\sigma_4(E_\epsilon, p)$, where p is the Fourier dual of the variable j so that the fourth order perturbation contribution vanishes. In this case σ acts as a convolution. However, this perturbative scheme is not convergent and must terminate at some order. We wish to estimate the remainder. This is where the real difficulty appears to lie. Consider the remainder term written in operator form:

$$< [G_0(E_\epsilon)\,\lambda v]^n G(E_\epsilon) > .$$

For brevity of notation we have omitted the contributions of the self-energy $\lambda^2\sigma$ which should also appear in this expression. Suppose we use the Schwarz inequality to separate $[G_0(E_\epsilon)\,\lambda v]^n$ from G. Then we have to estimate the expression:

$$< [G_0(E_\epsilon)\,\lambda v]^n \cdot [\bar{G}_0(E_\epsilon)\,\lambda v]^n > .$$

Here $\bar{G}_0$ denotes the complex conjugate of G_0. The high powers of λ may suggest that this is a small term. However, this is not so. If we write the above in matrix form we have

$$\lambda^{2n} < \sum_{j,k} [G_0(E_\epsilon; 0, j_1) v_{j_1} G_0(E_\epsilon; j_1, j_2) v_{j_2} \dots v_{j_{n-1}} G_0(E_\epsilon; j_{n-1}, j_n) v_{j_n}$$
$$\times \bar{G}_0(E_\epsilon; 0, k_1) v_{k_1} \bar{G}_0(E_\epsilon; k_1, k_2) v_{k_2} \dots v_{k_{n-1}} \bar{G}_0(E_\epsilon; k_{n-1}, j_n) v_{k_n}] > . \tag{4.119}$$

When computing the average over v the indices $\{j, k\}$ must be paired otherwise the expectation vanishes since $< v_j >= 0$ for each j. There are $n!$ such pairings, each of which gives rise to a graph. The pairing of any adjacent j_i, j_{i+1} or k_i, k_{i+1} is canceled by the self-energy contributions which we have omitted. We shall next discuss some other simple pairings.

After canceling the self-energy, the leading contribution to (4.119) should be given by *ladder graphs* of order n. These are a special class of graphs obtained by setting $j_i = k_i$ and summing over the vertices. Assuming G_0 is translation invariant, this sum is approximately given by:

$$[\lambda^2 \sum_j |G_0(E_\epsilon; 0, j)|^2]^n = [\frac{\lambda^2 Im\sigma(E_\epsilon)}{\lambda^2 Im\sigma(E_\epsilon) + \epsilon}]^n = [1 + \frac{\epsilon}{\lambda^2\, Im\sigma(E_\epsilon)}]^{-n}.$$

The right hand side of this equation is obtained using (4.118). Note that the right hand side of this equation is not small if $\epsilon \le \lambda^p$ unless $n \gg \lambda^{-(p-2)}$. Although contributions obtained from other pairings give smaller contributions, there are about $n!$ terms of size λ^{2n}. Hence we must require $n \ll \lambda^{-2}$ so that the number of graphs is offset by their size $n!\lambda^{2n} \ll 1$. Thus, $p < 4$ and this naive method of estimating $G(E_\epsilon)$ can only work for $\epsilon \gg \lambda^4$.

4.11.1 Leading Contribution to Diffusion

In three dimensions, the leading contribution to $< |G(E_\epsilon; 0, x)|^2 >$ is expected to be given by the sum of ladder graphs described above - denoted L. Let

$$K(E_\epsilon; p) = \sum_j e^{i\, j \cdot p} |G_0(E_\epsilon; 0, j)|^2.$$

Note that $K(p)$ is analytic for small values of p^2 since $Im\sigma > 0$. Then the sum of the ladders is given for $p \approx 0$ by

$$\hat{L}(p) = K(p)(1 - \lambda^2 K(p))^{-1} \approx \frac{K(0)}{1 - \lambda^2 K(0) - \frac{1}{2}\lambda^2 K''(0) p^2}$$
$$\approx \frac{Im\sigma}{\lambda^4\, Im\sigma\, K''(0)\, p^2/2 + \epsilon}.$$

In the last approximation we have used the Ward identity (4.118) to get

$$1 - \lambda^2 K(0) = \frac{\epsilon}{\lambda^2 Im\sigma + \epsilon} \quad so \quad K(0) \approx \lambda^{-2}.$$

Thus the sum of ladder graphs produces quantum diffusion as defined by (4.3) with $D_0 = \lambda^4 Im\sigma\, \hat{K}''(0)/2$. Since $K''(0) \propto \lambda^{-6}$, we see that D_0 is proportional to λ^{-2} for small λ. We refer to the work of Erdős, Salmhofer and Yau [21, 22], where such estimates were proved in 3D for $\epsilon \approx \lambda^{2+\delta}$, $\delta > 0$. At higher orders of perturbation theory, graphs in 3D must be grouped together to exhibit cancellations using Ward identities so that the bare diffusion propagator $1/p^2$ does not produce bogus divergences of the form $\int (1/p^2)^m$ for $m \geq 2$. These higher powers of are offset by a vanishing at p = 0 of the vertex joining such propagators. In both 1D and 2D this perturbation theory breaks down at higher orders. There is a divergent sum of crossed ladder graphs obtained by setting $j_i = k_{n-i}$ above. The sum of such graphs produces an integral of the form

$$\int_{|p| \leq 1} (p^2 + \epsilon)^{-1}\, d^2 p$$

which diverges logarithmically in 2D as $\epsilon \to 0$. This divergence is closely related to localization which is expected for small disorder in 2D [2].

4.11.2 Perturbation Theory for Random Band Matrices

To conclude this appendix, we briefly show how to adapt the perturbative methods described above to the case of random band matrices. Let H be a Gaussian random band matrix with covariance as in (4.14). Let

$$G(E_\epsilon) = (E_\epsilon - \lambda^2\sigma - \lambda H + \lambda^2\sigma)^{-1}.$$

Here the parameter λ is a book keeping device and it will be set equal to 1. Define $G_0(E_\epsilon) = [E_\epsilon - \lambda^2\sigma(E_\epsilon)]^{-1}$ and perturb $G(E_\epsilon)$ about G_0. Note that in this case G_0 is a scalar multiple of the identity. Proceeding as above we require that to second order perturbation theory in λ vanishes. From (4.13) and (4.14) we have

$$\sum_k < H_{jk}H_{kj} >= \sum_k J_{jk} = 1$$

and we obtain the equation

$$G_0 = [E_\epsilon - \lambda^2\sigma(E_\epsilon)]^{-1} = \sigma(E_\epsilon).$$

This is a quadratic equation for σ and when we set $\lambda = 1$ and take the imaginary part we recover Wigner's density of states. For a band matrix of width W the corrections at λ^4 to the Wigner semicircle law is formally of order $1/W^2$. If apply the Schwarz inequality as above we will encounter similar difficulties. In fact it is known that $\epsilon \gg W^{-2/5}$ is needed to ensure convergence of perturbation theory. However, by using different methods, Erdős et al. [23] and Sodin [44] establish control of the DOS to scale $\epsilon \approx W^{-1}$. For 3D Gaussian band matrices with covariance given by (4.14) these difficulties are avoided with nonperturbative SUSY methods [7, 13] and ϵ may be sent to zero for fixed large W. The calculation of bare diffusion constant explained above for random Schrödinger matrices can be applied to RBM. In this case $D_0 \approx W^2$.

We now illustrate the perturbation theory for the two site N-orbital model:

$$M = \begin{pmatrix} \lambda A & I \\ I & \lambda B \end{pmatrix}$$

where A and B are independent $N \times N$ GUE matrices and I is the $N \times N$ identity matrix. Define

$$G_0^{-1} = \begin{pmatrix} E_\epsilon - \lambda^2\sigma & I \\ I & E_\epsilon - \lambda^2\sigma \end{pmatrix}.$$

By requiring second order perturbation theory to vanish as above we obtain a cubic equation for the self-energy $\sigma = (E_\epsilon - \lambda^2\sigma)[(E_\epsilon - \lambda^2\sigma)^2 - 1]^{-1}$. This equation is analogous to (4.117). The imaginary part of $\sigma(E)$ is proportional to the density of states for M for large N. To calculate finer properties of the density of states one can apply the SUSY methods of Sects. 4.3–4.5 and it can be shown that the main saddle point will coincide the self-energy σ. Note that the density of states for M no longer looks like the semicircle law, in fact it is peaked near ± 1 when λ is small.

Nevertheless, the local eigenvalue spacing distribution is expected to be equal to that of GUE after a local rescaling.

Remark. One of the major challenges in the analysis of random band or random Schrödinger matrices is to find a variant of perturbation theory which allows one to analyze smaller values of ϵ as a function of W or λ. For certain Gaussian band matrices SUSY gives much better estimates but the class of models to which it applies is relatively narrow. Perhaps there is a method which combines perturbative and SUSY ideas and emulates both of them.

4.12 Appendix C: Bounds on Green's Functions of Divergence Form

In this appendix we will show how to obtain upper and lower bounds on (4.87)

$$< [fe^t;\ [D_{\beta,\varepsilon}(t)]^{-1} fe^t] >= \sum_{x,y} \mathcal{G}_{\beta,\varepsilon}(x,y)\, f(x) f(y)$$

in terms of $[f; G_0 f] = [f; (\beta\nabla^*\nabla + \epsilon)^{-1} f]$ for $f(j) \geq 0$. Recall [;] is the scalar product and that $D_{\beta,\varepsilon}(t)$ is given by (4.71) or (4.77). We shall suppose that $d \geq 3$ and that $< cosh^8(t_j) >$ is bounded as in Theorem 14 of Sect. 4.8. For brevity let

$$G_t(i,j) = [D_{\beta,\varepsilon}(t)]^{-1}(i,j).$$

We first prove the lower bound. For any two real vectors X and Y we have the inequality

$$X \cdot X \geq 2X \cdot Y - Y \cdot Y.$$

Let

$$X_1(j,\alpha) = \sum_k e^{(t_j + t_{j+e_\alpha})/2} \nabla_\alpha G_t(j,k) f(k) e^{t_k}$$

where e_α is the unit vector in the α direction. Define

$$X_2(j) = \sqrt{\epsilon}\, e^{t_j/2} \sum_k G_t(j,k) f(k) e^{t_k}.$$

If we set $X = (X_1, X_2)$ we see that $< X \cdot X >= [f; \mathcal{G}_{\beta,\varepsilon} f]$. We shall define Y to be proportional to X with G_t replaced by G_0,

$$Y_1(j,\alpha) = a\, e^{-2t_0} \sum_k e^{(t_j + t_{j+e_\alpha})/2} \nabla_\alpha G_0(j,k) f(k) e^{t_k}$$

and

$$Y_2(j) = a\, e^{-2t_0} \sqrt{\epsilon}\, e^{t_j/2} \sum_k G_0(j,k) f(k) e^{t_k}.$$

The constant a is chosen so that the error term $Y \cdot Y$ is small. By integrating by parts we have

$$X \cdot Y = a\, [f e^t;\ G_0 f e^t] e^{-2t_0}.$$

Since $< e^{t_k + t_j - 2t_0} > \geq 1$ by Jensen's inequality and translation invariance, we get the desired lower bound on $< X \cdot Y >$ in terms of $a\, [f; G_0 f]$. The error term

$$< Y_1 \cdot Y_1 > = a^2 \ < [\nabla G_0 f e^t;\ e^{t_j + t_{j'}} \nabla G_0 f e^t]\, e^{-4t_0} > \leq C\, a^2 [f; G_0 f].$$

The last inequality follows from our bounds on $< \cosh^4(t_j - t_k) >$ which can be bounded by $< \cosh^8 t_j >$. Note that we also need to use $f \geq 0$ and

$$\sum_{ijk} f(i) |\nabla G_0(i,j)|\, |\nabla G_0(j,k)| f(k) \leq Const\, [f;\ G_0 f] \tag{4.120}$$

which holds in three dimensions.

$$< Y_2 \cdot Y_2 > = a^2 \epsilon < [e^t G_0 e^t f;\ G_0 e^t f] e^{-4t_0} > \leq C\, a^2\, [f;\ G_0 f].$$

In the last inequality we bound $< e^{-t_0} >$ as well as $< cosh^4(t_j - t_k) >$. The parameter a is chosen small so that $< X \cdot Y >$ is the dominant term.

The upper bound is a standard estimate obtained as follows. Let $L_0 = \beta \nabla^* \nabla + \varepsilon$ and $G_0 = L_0^{-1}$. Then

$$[f e^t;\ G_t f e^t] = [L_0 G_0\, f e^t;\ G_t\, f e^t].$$

Integrating by parts we see that the right side equals

$$[\sqrt{\beta} e^{-(t_j + t_{j'})/2} \nabla G_0\, f e^t;\ \sqrt{\beta} e^{(t_j + t_{j'})/2} \nabla G_t\, f e^t] + [\sqrt{\varepsilon} e^{-t/2} G_0 f e^t;\ \sqrt{\varepsilon} e^{t/2} G_t\, f e^t].$$

By the Schwarz inequality we get

$$[f e^t; G_t f e^t]$$

$$\leq [f e^t; G_t f e^t]^{1/2} \left(\beta \sum_j |\nabla_j (G_0 f e^t)|^2 e^{-(t_j + t_{j'})} + \varepsilon \sum_j |(G_0 f e^t)(j)|^2 e^{-t_j} \right)^{1/2}$$

Therefore

$$[fe^t; G_t fe^t] \leq \beta \sum_j |\nabla_j (G_0 e^t f)|^2 e^{-(t_j+t_{j'})} + \varepsilon \sum_j |(G_0 e^t f)(j)|^2 e^{-t_j}. \tag{4.121}$$

The proof of the upper bound now follows using (4.120). Note that in both inequalities we need to assume that $f \geq 0$.

The desired upper bound on the expectation (4.121) in terms of G_0 now follows from bounds on $< cosh^8(t_j) >$ and (4.120). Note that if the factor of e^t times f were not present we would not need to require $f \leq 0$ and to use (4.120).

Acknowledgements I wish to thank my collaborators Margherita Disertori and Martin Zirnbauer for making this review on supersymmetry possible. Thanks also to Margherita Disertori and Sasha Sodin for numerous helpful comments and corrections on early versions of this review and to Yves Capdeboscq for improving the lower bound in (4.87). I am most grateful to the organizers of the C.I.M.E. summer school, Alessandro Giuliani, Vieri Mastropietro, and Jacob Yngvason for inviting me to give these lectures and for creating a stimulating mathematical environment. Finally, I thank Abdelmalek Abdesselam for many helpful questions during the course of these lectures.

References

1. A. Abdesselam, Grassmann-Berezin calculus and theorems of the matrix-tree type. Adv. Appl. Math. **33**, 51–70 (2004). http://arxiv.org/abs/math/0306396
2. E. Abrahams, P.W. Anderson, D.C. Licciardello, T.V. Ramakrishnan, Scaling theory of localization: Absence of quantum diffusion in two dimensions. Phys. Rev. Lett. **42**, 673 (1979)
3. E.J. Beamond, A.L. Owczarek, J. Cardy, Quantum and classical localisation and the Manhattan lattice. J. Phys. A Math. Gen. **36**, 10251 (2003) [arXiv:cond-mat/0210359]
4. F.A. Berezin, *Introduction to Superanalysis* (Reidel, Dordrecht, 1987)
5. H. Brascamp, E. Lieb, On extensions of the Brunn-Minkowski and Prekopa-Leindler theorems, including inequalities for log concave functions, and with an application to the diffusion equation. J. Func. Anal. **22**, 366–389 (1976)
6. E. Brezin, V. Kazakov, D. Serban, P. Wiegman, A. Zabrodin, in *Applications of Random Matrices to Physics*. Nato Science Series, vol. 221 (Springer, Berlin, 2006)
7. F. Constantinescu, G. Felder, K. Gawedzki, A. Kupiainen, Analyticity of density of states in a gauge-invariant model for disordered electronic systems. J. Stat. Phys. **48**, 365 (1987)
8. P. Deift, *Orthogonal Polynomials, and Random Matrices: A Riemann-Hilbert Approach* (CIMS, New York University, New York, 1999)
9. P. Deift, T. Kriecherbauer, K.T.-K. McLaughlin, S. Venakides, X. Zhou, Uniform asymptotics for polynomials orthogonal with respect to varying exponential weights and applications to universality questions in random matrix theory. Comm. Pure Appl. Math. **52**, 1335–1425 (1999)
10. P. Diaconis, in *Recent Progress on de Finetti's Notions of Exchangeability*. Bayesian Statistics, vol. 3 (Oxford University Press, New York, 1988), pp. 111–125
11. M. Disertori, Density of states for GUE through supersymmetric approach. Rev. Math. Phys. **16**, 1191–1225 (2004)
12. M. Disertori, T. Spencer, Anderson localization for a SUSY sigma model. Comm. Math. Phys. **300**, 659–671 (2010). http://arxiv.org/abs/0910.3325
13. M. Disertori, H. Pinson T. Spencer, Density of states for random band matrices. Comm. Math. Phys. **232**, 83–124 (2002). http://arxiv.org/abs/math-ph/0111047
14. M. Disertori, T. Spencer, M.R. Zirnbauer, Quasi-diffusion in a 3D supersymmetric hyperbolic sigma model. Comm. Math Phys. **300**, 435 (2010). http://arxiv.org/abs/0901.1652

15. K.B. Efetov, in *Anderson Localization and Supersymmetry*, ed. by E. Abrahams. 50 Years of Anderson Localization (World Scientific, Singapore, 2010). http://arxiv.org/abs/1002.2632
16. K.B. Efetov, Minimum metalic conductivity in the theory of localization. JETP Lett. **40**, 738 (1984)
17. K.B. Efetov, Supersymmetry and theory of disordered metals. Adv. Phys. **32**, 874 (1983)
18. K.B. Efetov, *Supersymmetry in Disorder and Chaos* (Cambridge University Press, Cambridge, 1997)
19. L. Erdős, Universality of Wigner random matrices: A survey of recent results. Russian Math. Surveys. **66**(3) 507–626 (2011). http://arxiv.org/abs/1004.0861
20. L. Erdős, B. Schlein, H-T. Yau, Local semicircle law and complete delocalization for Wigner random matrices. Comm. Math. Phys. **287**, 641–655 (2009)
21. L. Erdős, M. Salmhofer, H-T. Yau, Quantum diffusion of the random Schrödinger evolution in the scaling limit. Acta Math. **200**, 211 (2008)
22. L. Erdős, in Les Houches Lectures: *Quantum Theory from Small to Large Scales*, Lecture notes on quantum Brownian motion, vol. 95, http://arxiv.org/abs/1009.0843
23. L. Erdős, H.T. Yau, J. Yin, Bulk universality for generalized Wigner matrices. to appear in Prob. Th. and Rel.Fields. [arxiv:1001.3453]
24. J. Feldman, H. Knörrer, E. Trubowitz, in *Fermionic Functional Integrals and the Renormalization Group*. CRM Monograph Series (AMS, Providence, 2002)
25. J. Fröhlich, T. Spencer, The Kosterlitz Thouless transition. Comm. Math. Phys. **81**, 527 (1981)
26. J. Fröhlich, T. Spencer, On the statistical mechanics of classical Coulomb and dipole gases. J. Stat. Phys. **24**, 617–701 (1981)
27. J. Fröhlich, B. Simon, T. Spencer, Infrared bounds, phase transitions and continuous symmetry breaking. Comm. Math. Phys. **50**, 79 (1976)
28. Y.V. Fyodorov, in *Basic Features of Efetov's SUSY*, ed. by E. Akkermans et al. Mesoscopic Quantum Physics (Les Houches, France, 1994)
29. Y.V. Fyodorov, Negative moments of characteristic polynomials of random matrices. Nucl. Phys. B **621**, 643–674 (2002). http://arXiv.org/abs/math-ph/0106006
30. I.A. Gruzberg, A.W.W. Ludwig, N. Read, Exact exponents for the spin quantun Hall transition. Phys. Rev. Lett. **82**, 4524–4527 (1999)
31. T. Guhr, in *Supersymmetry in Random Matrix Theory*. The Oxford Handbook of Random Matrix Theory (Oxford University Press, Oxford, 2010). http://arXiv:1005.0979v1
32. J.M. Kosterlitz, D.J. Thouless, Ordering, metastability and phase transitions in two-dimensional systems. J. Phys. C **6**, 1181 (1973)
33. P. Littelmann, H.-J. Sommers, M.R. Zirnbauer, Superbosonization of invariant random matrix ensembles. Comm. Math. Phys. **283**, 343–395 (2008)
34. F. Merkl, S. Rolles, in *Linearly Edge-Reinforced Random Walks*. Dynamics and Stochastics. Lecture Notes-Monograph Series, vol. 48 (2006), pp. 66–77
35. F. Merkl, S. Rolles, Edge-reinforced random walk on one-dimensional periodic graphs. Probab. Theor. Relat. Field **145**, 323 (2009)
36. A.D. Mirlin, in *Statistics of Energy Levels*, ed. by G.Casati, I. Guarneri, U. Smilansky. New Directions in Quantum Chaos. Proceedings of the International School of Physics "Enrico Fermi", Course CXLIII (IOS Press, Amsterdam, 2000), pp. 223–298. http://arXiv.org/abs/cond-mat/0006421
37. L. Pastur, M. Shcherbina, Universality of the local eigenvalue statistics for a class of unitary invariant matrix ensembles. J. Stat. Phys. **86**, 109–147 (1997)
38. R. Peamantle, Phase transition in reinforced random walk and RWRE on trees. Ann. Probab. **16**, 1229–1241 (1988)
39. Z. Rudnick, P. Sarnak, Zeros of principal L-functions and random-matrix theory. Duke Math. J. **81**, 269–322 (1996)
40. M. Salmhofer, *Renormalization: An Introduction* (Springer, Berlin, 1999)
41. L. Schäfer, F. Wegner, Disordered system with n orbitals per site: Lagrange formulation, hyperbolic symmetry, and Goldstone modes. Z. Phys. B **38**, 113–126 (1980)

42. J. Schenker, Eigenvector localization for random band matrices with power law band width. Comm. Math. Phys. **290**, 1065–1097 (2009)
43. A. Sodin, The spectral edge of some random band matrices. Ann. Math. **172**, 2223 (2010)
44. A. Sodin, An estimate on the average spectral measure for random band matrices, J. Stat. Phys. **144**, 46–59 (2011). http://arxiv.org/1101.4413
45. T. Spencer, in *Mathematical Aspects of Anderson Localization*, ed. by E. Abrahams. 50 Years of Anderson Localization (World Scientific, Singapore, 2010)
46. T. Spencer, in *Random Band and Sparse Matrices*. The Oxford Handbook of Random Matrix Theory, eds. by G. Akermann, J. Baik, and P. Di Francesco. (Oxford University Press, Oxford, 2010)
47. T. Spencer, M.R. Zirnbauer, Spontaneous symmetry breaking of a hyperbolic sigma model in three dimensions. Comm. Math. Phys. **252**, 167–187 (2004). http://arXiv.org/abs/math-ph/0410032
48. B. Valko, B. Virag, Random Schrödinger operators on long boxes, noise explosion and the GOE, http://arXiv.org/abs/math-ph/09120097
49. J.J.M. Verbaarschot, H.A. Weidenmüller, M.R. Zirnbauer, Phys. Rep. **129**, 367–438 (1985)
50. F. Wegner, The Mobility edge problem: Continuous symmetry and a Conjecture. Z. Phys. B **35**, 207–210 (1979)
51. M.R. Zirnbauer, The Supersymmetry method of random matrix theory, http://arXiv.org/abs/math-ph/0404057(2004)
52. M.R. Zirnbauer, Fourier analysis on a hyperbolic supermanifold with constant curvature. Comm. Math. Phys. **141**, 503–522 (1991)
53. M.R. Zirnbauer, Riemannian symmetric superspaces and their origin in random-matrix theory. J. Math. Phys. **37**, 4986–5018 (1996)

[illegible]

List of Participants

1. Aaen Anders
   aaen@imf.au.dk
2. Abdesselam Abdelmalek
   malek@virginia.edu
3. Adami Riccardo
   riccardo.adami@unimib.it
4. Alazzawi Sabina
   nabil_alazzawi@gmx.at
5. Ballesteros Miguel
   ballesteros.miguel.math@gmail.com
6. Bölzle Sebastian
   sebo@fa.uni-tuebingen.de
7. Cenatiempo Serena
   Serena.Cenatiempo@roma1.infn.it
8. Chandra Ajay
   ac2yx@virginia.edu
9. Correggi Michele
   michele.correggi@gmail.com
10. Costa Emanuele
    ecosta@sissa.it
11. De Nittis Giuseppe
    denittis@sissa.it
12. Draxler Damian
    damian.draxler@gmx.net
13. Finco Domenico
    finco@mat.uniroma1.it
14. Giuliani Alessandro *(editor)*
    giuliani@mat.uniroma3.it
15. Greenblatt Rafael
    greenbla@mat.uniroma3.it

V. Rivasseau et al., *Quantum Many Body Systems*, Lecture Notes in Mathematics 2051,
DOI 10.1007/978-3-642-29511-9, © Springer-Verlag Berlin Heidelberg 2012

16. Kirkpatrik Kirk
kirkpatr@cims.nyu.edu
17. Kopsky Georg
georg.kopsky@univie.ac.at
18. Kuna Tobias
t.kuna@reading.ac.uk
19. Marchesiello Antonella
marchesiello@dmmm.uniroma1.it
20. Mastropietro Vieri *(editor)*
mastropi@mat.uniroma2.it
21. Nam Phan Thanh
ptnam@math.ku.dk
22. Ortoleva Cecilia
c.ortoleva@campus.unimib.it
23. Panati Gianluca
panati@mat.uniroma1.it
24. Pinsker Florian
florian.pinsker@gmail.com
25. Porta Marcello
marcello.porta@roma1.infn.it
26. Rivasseau Vincent *(lecturer)*
vincent.rivasseau@gmail.com
27. Seiringer Robert *(lecturer)*
rseiring@math.mcgill.ca
28. Simonella Sergio
simonella@mat.uniroma1.it
29. Smerlak Matteo
matteo.smerlak@gmail.com
30. Solovej Jan Philip *(lecturer)*
solovey@math.ku.dk
31. Spencer Thomas *(lecturer)*
spencer@math.ias.edu
32. Tsvir Zhanna
janna_cvir@mail.ru
33. Ueltschi Daniel
daniel@ueltschi.org
34. von Keler Johannes
jovo@fa.uni-tuebingen.de
35. Windridge Peter
p.windridge@warwick.ac.uk
36. Yngvason Jakob *(editor)*
jakob.yngvason@univie.ac.at

LECTURE NOTES IN MATHEMATICS

Springer

Edited by J.-M. Morel, B. Teissier; P.K. Maini

Editorial Policy (for Multi-Author Publications: Summer Schools / Intensive Courses)

1. Lecture Notes aim to report new developments in all areas of mathematics and their applications - quickly, informally and at a high level. Mathematical texts analysing new developments in modelling and numerical simulation are welcome. Manuscripts should be reasonably selfcontained and rounded off. Thus they may, and often will, present not only results of the author but also related work by other people. They should provide sufficient motivation, examples and applications. There should also be an introduction making the text comprehensible to a wider audience. This clearly distinguishes Lecture Notes from journal articles or technical reports which normally are very concise. Articles intended for a journal but too long to be accepted by most journals, usually do not have this "lecture notes" character.

2. In general SUMMER SCHOOLS and other similar INTENSIVE COURSES are held to present mathematical topics that are close to the frontiers of recent research to an audience at the beginning or intermediate graduate level, who may want to continue with this area of work, for a thesis or later. This makes demands on the didactic aspects of the presentation. Because the subjects of such schools are advanced, there often exists no textbook, and so ideally, the publication resulting from such a school could be a first approximation to such a textbook. Usually several authors are involved in the writing, so it is not always simple to obtain a unified approach to the presentation.
   For prospective publication in LNM, the resulting manuscript should not be just a collection of course notes, each of which has been developed by an individual author with little or no coordination with the others, and with little or no common concept. The subject matter should dictate the structure of the book, and the authorship of each part or chapter should take secondary importance. Of course the choice of authors is crucial to the quality of the material at the school and in the book, and the intention here is not to belittle their impact, but simply to say that the book should be planned to be written by these authors jointly, and not just assembled as a result of what these authors happen to submit.
   This represents considerable preparatory work (as it is imperative to ensure that the authors know these criteria before they invest work on a manuscript), and also considerable editing work afterwards, to get the book into final shape. Still it is the form that holds the most promise of a successful book that will be used by its intended audience, rather than yet another volume of proceedings for the library shelf.

3. Manuscripts should be submitted either online at www.editorialmanager.com/lnm/ to Springer's mathematics editorial, or to one of the series editors. Volume editors are expected to arrange for the refereeing, to the usual scientific standards, of the individual contributions. If the resulting reports can be forwarded to us (series editors or Springer) this is very helpful. If no reports are forwarded or if other questions remain unclear in respect of homogeneity etc, the series editors may wish to consult external referees for an overall evaluation of the volume. A final decision to publish can be made only on the basis of the complete manuscript; however a preliminary decision can be based on a pre-final or incomplete manuscript. The strict minimum amount of material that will be considered should include a detailed outline describing the planned contents of each chapter.
   Volume editors and authors should be aware that incomplete or insufficiently close to final manuscripts almost always result in longer evaluation times. They should also be aware that parallel submission of their manuscript to another publisher while under consideration for LNM will in general lead to immediate rejection.

4. Manuscripts should in general be submitted in English. Final manuscripts should contain at least 100 pages of mathematical text and should always include
   - a general table of contents;
   - an informative introduction, with adequate motivation and perhaps some historical remarks: it should be accessible to a reader not intimately familiar with the topic treated;
   - a global subject index: as a rule this is genuinely helpful for the reader.

   Lecture Notes volumes are, as a rule, printed digitally from the authors' files. We strongly recommend that all contributions in a volume be written in the same LaTeX version, preferably LaTeX2e. To ensure best results, authors are asked to use the LaTeX2e style files available from Springer's web-server at
   ftp://ftp.springer.de/pub/tex/latex/svmonot1/ (for monographs) and
   ftp://ftp.springer.de/pub/tex/latex/svmultt1/ (for summer schools/tutorials).
   Additional technical instructions, if necessary, are available on request from:
   lnm@springer.com.

5. Careful preparation of the manuscripts will help keep production time short besides ensuring satisfactory appearance of the finished book in print and online. After acceptance of the manuscript authors will be asked to prepare the final LaTeX source files and also the corresponding dvi-, pdf- or zipped ps-file. The LaTeX source files are essential for producing the full-text online version of the book. For the existing online volumes of LNM see:
   http://www.springerlink.com/openurl.asp?genre=journal&issn=0075-8434.
   The actual production of a Lecture Notes volume takes approximately 12 weeks.

6. Volume editors receive a total of 50 free copies of their volume to be shared with the authors, but no royalties. They and the authors are entitled to a discount of 33.3 % on the price of Springer books purchased for their personal use, if ordering directly from Springer.

7. Commitment to publish is made by letter of intent rather than by signing a formal contract. Springer-Verlag secures the copyright for each volume. Authors are free to reuse material contained in their LNM volumes in later publications: a brief written (or e-mail) request for formal permission is sufficient.

Addresses:

Professor J.-M. Morel, CMLA,
École Normale Supérieure de Cachan,
61 Avenue du Président Wilson, 94235 Cachan Cedex, France
E-mail: morel@cmla.ens-cachan.fr

Professor B. Teissier, Institut Mathématique de Jussieu,
UMR 7586 du CNRS, Équipe "Géométrie et Dynamique",
175 rue du Chevaleret,
75013 Paris, France
E-mail: teissier@math.jussieu.fr

For the "Mathematical Biosciences Subseries" of LNM:

Professor P. K. Maini, Center for Mathematical Biology,
Mathematical Institute, 24-29 St Giles,
Oxford OX1 3LP, UK
E-mail : maini@maths.ox.ac.uk

Springer, Mathematics Editorial I,
Tiergartenstr. 17,
69121 Heidelberg, Germany,
Tel.: +49 (6221) 4876-8259
Fax: +49 (6221) 4876-8259
E-mail: lnm@springer.com